Fixed Point Theory and Chaos

Edited by Guillermo Huerta-Cuellar

Published in London, United Kingdom

Fixed Point Theory and Chaos
http://dx.doi.org/10.5772/intechopen.104025
Edited by Guillermo Huerta-Cuellar

Contributors
Getahun Bekele Wega, Mohamad Muslikh, Adem Kilicman, Jamnian Nantadilok, Buraskorn Nuntadilok, Tayebe Lal Shateri, Konstantinos Tzirakis, Guillermo Huerta-Cuellar, Hafiz Muhammad Zeeshan

First published in London, United Kingdom, 2023 by IntechOpen
IntechOpen is the global imprint of INTECHOPEN LIMITED, registered in England and Wales, registration number: 11086078, 5 Princes Gate Court, London, SW7 2QJ, United Kingdom

British Library Cataloguing-in-Publication Data
A catalogue record for this book is available from the British Library

Additional hard and PDF copies can be obtained from orders@intechopen.com

Fixed Point Theory and Chaos
Edited by Guillermo Huerta-Cuellar
p. cm.
Print ISBN 978-1-83768-435-9
Online ISBN 978-1-83768-436-6
eBook (PDF) ISBN 978-1-83768-437-3

Meet the editor

Guillermo Huerta Cuellar earned his BSc degree from Instituto de Investigación en Comunicaciones Ópticas (IICO), UASLP in 2004 and his Ph.D. from Centro de Investigaciones en Óptica (CIO) in 2009. He has been working at Centro Universitario de los Lagos, University of Guadalajara, México, since 2010. During this time, he also served as a visiting researcher at the Department of Applied Mathematics IPICYT, México (2012-2014), the Faculty of Radiophysics, Lobachevsky State University of Nizhny Novgorod, Russia (2016), and had sabbaticals at St. Mary's University, San Antonio, Texas, USA (2018-2019), and IPICYT, México (2019-2020). He has edited three books, authored seven book chapters, and published more than 70 high-impact papers. Since 2019, he has co-organized the International Meeting for Dissemination and Research in the Study of Complex Systems and their Applications (EDIESCA), which is held annually in several Mexican universities. His research interests include the study, characterization, dynamical behavior, and design of nonlinear dynamical systems such as lasers, electronics, and numerical models.

Contents

Preface

The fixed point theory is a crucial area of study in both theoretical and applied mathematics. Its applications can be seen in various fields such as physics, chemistry, and economics, among others. In the realm of mathematics, fixed point theory finds applications in differential equations, game theory, and integral theory equations. This theory has been extensively used to solve integral equations of first-order differential equations for linear, nonlinear, or chaotic systems.

The current book presents recent results on the study of fixed points with different perspectives. The introductory chapter covers the basics of chaos and fixed points. Chapter 2 focuses on coupled fixed points for (ϕ, ψ)-contractive mappings in partially ordered modular spaces, where the Banach contraction principle is one of the primary tools used to study the fixed points of contractive maps in the framework of modular space endowed with a partial order. Chapter 3 discusses common fixed points of asymptotically quasi-nonexpansive mappings in CAT(0) spaces. Chapter 4 is devoted to the study of iterative algorithms for common solutions of nonlinear problems in Banach spaces. Chapter 5 explains fixed points for the derivative of set-valued functions. Chapter 6 examines stability estimates for fractional Hardy-Schrödinger operators and derives Hardy-Sobolev-type improvements in fractional Hardy inequalities.

Guillermo Huerta-Cuellar
Centro Universitario de los Lagos,
University of Guadalajara,
Exact Sciences and Technology Department,
Guadalajara, Mexico

Section 1

Introduction

Chapter 1

Introductory Chapter: Fixed Points Theory and Chaos

Guillermo Huerta-Cuellar and Hafiz Muhammad Zeeshan

1. Introduction

Among the systems that exhibit dynamical behavior, nonlinear and chaotic systems are the most intriguing, as they exhibit an enormous variety of performances and offer a great opportunity for technological applications. The formal study of chaotic systems begins with the results reported by Lorenz [1]. In this sense, the study and characterization of dynamical systems, especially chaotic systems, is one of the breakthroughs of the last century, although it is a relatively new field of research that is becoming increasingly important in various scientific disciplines [2–6]. In the case of nonlinear maps, it has been found that chaos can also arise between the dynamic behavior that these maps produce [7–9]. The study of fixed points could prepare the scientific community to investigate how to stabilize the behavior of multiple dynamical systems that generally exhibit nonlinear behavior, which is of great importance in current issues [10, 11]. The stabilization of fixed points in chaotic systems is one of the most interesting topics in the study of systems with chaotic behavior. Among the systems that have been stabilized are Lorenz, Rössler, and Chua [12–14]. As mentioned earlier, there are many works in which the chaotic behavior can be controlled by stabilizing the system's fixed points. However, it is also possible to control the stabilization of the fixed points to obtain stable or multistable behavior of chaotic systems [15–17]. Moreover, this behavior has been studied in both integer and fractional-order systems [18]. Recently, Echenausía-Monroy et al. [19] presented an interesting method to characterize qualitative changes in the dynamical behavior of a family of piecewise linear systems by controlling the transition from monostable to multistable oscillations around different fixed points by studying the stable and unstable manifolds and their relation to the eigendirections.

2. A brief definition of fixed points

In the field of applied mathematics, fixed-point theory refers to an interdisciplinary topic that can be applied in various disciplines like economics, variational inequalities, approximation theory, game theory, and optimization theory, among other areas of interest. Fixed-point theory is divided into three major areas, as can be seen in **Figure 1**.

Topological fixed-point theory was developed by L.E.J. Brouwer in 1912 [20]. According to Brouwer "Every continuous function from convex compact subset K of a Euclidian space to K itself has a fixed point." It has several real-world illustrations.

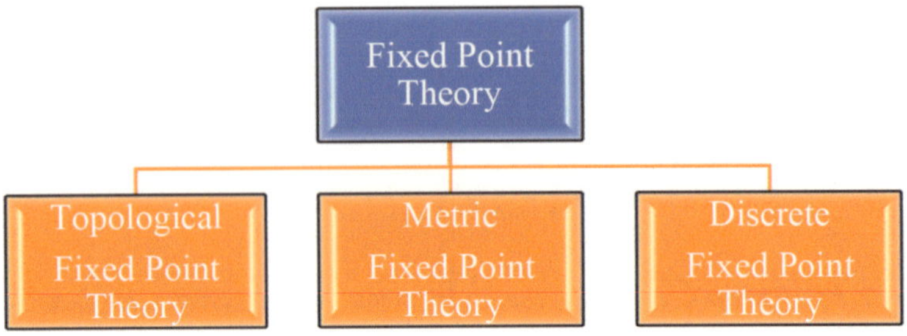

Figure 1.
Fixed points components.

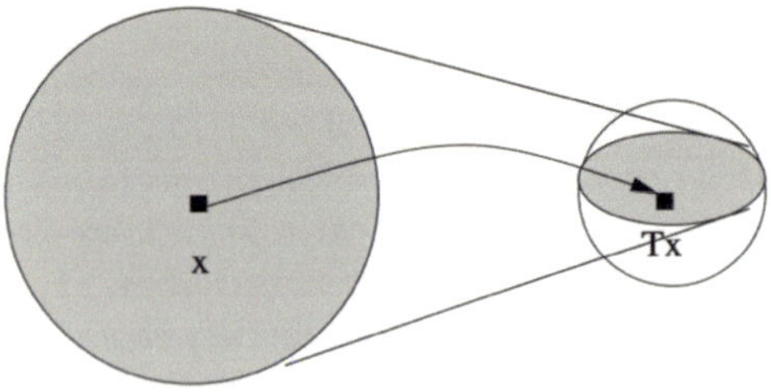

Figure 2.
Contraction mapping [23].

Consider a map of the country. If this map were placed anywhere in that country, there would always be a point on the map representing exactly that point.

One of the pioneering works about fixed points is Henri Poincaré [21], which was proposed as the first work about fixed points in 1886. Although the basic concept of *metric fixed-point theory* was known to others previously, the Polish mathematician Stefan Banach is credited with making it usable and well-known. The Banach *Fixed Point Theorem* (also known as the contraction mapping theorem or contraction mapping principle) is a useful tool in the study of metric spaces. It ensures the presence and uniqueness of fixed points of particular self-maps of metric spaces and gives a constructive approach to finding such fixed points [22]. The theorem is named after Stefan Banach (1892–1945) and was first stated by him in 1922. Banach stated that "Let (X, d) be a metric space." A mapping $T : X \to X$ is called Banach contraction mapping if there exists a constant $k \in [0,1)$ (s.t)

$$d(Tx,Ty) \le k.d(x,y) \; forall x, y \in X \tag{1}$$

Some fixed points theorems and different spaces were from the study and generalization of fixed points, as well as the Banach contraction theorem (**Figure 2**).

The discrete fixed-point theory came from Alfred Traski in 1955. Traski proved that "If *F* is a monotone function on a nonempty complete lattice, then the set of fixed points of *F* forms a nonempty complete lattice" [24].

But what is a fixed point? In this sense a short and comprehensive definition and interesting example are given next:

Let X be a nonempty set and $T: X \rightarrow X$ be a mapping. Then $x \in X$ is known as fixed point of T if $Tx = x$

Graphically, these are the places at which the graph of f whose equation is $y = f(x)$, crosses the diagonal, whose equation $y = x$.

Let $y = f(x) = x^3 + 4x^2 - 3x - 16$, then it has three fixed points $x = 2, x = -2 \wedge x = -4$ as shown in **Figure 3**.

A fixed point is a location that stays the same when a map, set of differential equations, etc. are applied to it. Informally, the area of mathematics known as fixed point theory aims to locate all self-maps or self-correspondences in which at least one element is left invariant.

- Fixed Point for single-valued mapping

The fixed point for the mapping $S: R \rightarrow R$ defined as $S(x) = \frac{x}{2}$ is distinct. Obviously, the only fixed point is 0.

- Fixed Point for multi-valued mapping

a. There are two fixed points in the mapping $S: R \rightarrow R$ defined as $S(x) = \sqrt{x}$. The only fixed point, in this case, is 0 and 1.

b. There are infinitely many fixed points in the mapping $S: R^2 \rightarrow R^2$ defined as $T(x,y) = x$. In fact, all points of $x - axis$ are fixed points.

A Mapping may have a unique fixed point, more than one, or infinitely many fixed points.

Remark: There may exist mapping which not has a fixed point.

Example: Let X be a nonempty set. There is no fixed point in the mapping $S: X \rightarrow X$ defined as $S(x) = x + a$ where 'a' is any constant.

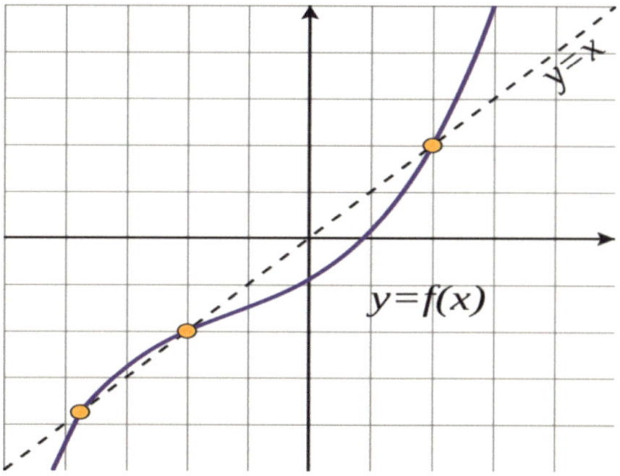

Figure 3.
Graphically representation of fixed point [25].

Author details

Guillermo Huerta-Cuellar[1*] and Hafiz Muhammad Zeeshan[1,2]

1 University of Guadalajara, Mexico

2 Divisional Public School and College, Rawalpindi, Pakistan

*Address all correspondence to: ghuertacuellar@gmail.com

References

[1] Lorenz EN. The problem of deducing the climate from the governing equations. Tellus. 1964;**16**(1):1-11

[2] Akhmet M, Yeşil C, Başkan K. Synchronization of chaos in semiconductor gas discharge model with local mean energy approximation. Chaos, Solitons & Fractals. 2023;**167**:113035

[3] Messadi M et al. A new 4D Memristor chaotic system: Analysis and implementation. Integration. 2023;**88**:91-100

[4] Echenausía-Monroy JL et al. Deterministic Brownian-like motion: Electronic approach. Electronics. 2022;**11**(18):2949

[5] Meucci R et al. Generalized multistability and its control in a laser. Chaos: An interdisciplinary. Journal of Nonlinear Science. 2022;**32**(8):083111

[6] Afanador Delgado SM et al. Implementation of logic gates in an erbium-doped fiber laser (EDFL): Numerical and experimental analysis. Photonics. 2022;**9**:977

[7] May RM. Simple mathematical models with very complicated dynamics. Nature. 1976;**261**:459-467

[8] Devaney R, Nitecki Z. Shift automorphisms in the Hénon mapping. Communications in Mathematical Physics. 1979;**67**:137-146

[9] Lozi R. Un attracteur étrange (?) du type attracteur de Hénon. Le Journal de Physique Colloques. 1978;**39**(C5):C5-9-C5-10

[10] Barak O et al. From fixed points to chaos: Three models of delayed discrimination. Progress in Neurobiology. 2013;**103**:214-222

[11] Schiff SJ et al. Controlling chaos in the brain. Nature. 1994;**370**(6491):615-620

[12] Yang S-K, Chen C-L, Yau H-T. Control of chaos in Lorenz system. Chaos, Solitons & Fractals. 2002;**13**(4):767-780

[13] Bodale I, Oancea VA. Chaos control for Willamowski–Rössler model of chemical reactions. Chaos, Solitons & Fractals. 2015;**78**:1-9

[14] Wu T, Chen M-S. Chaos control of the modified Chua's circuit system. Physica D: Nonlinear Phenomena. 2002;**164**(1-2):53-58

[15] Pisarchik AN, Feudel U. Control of multistability. Physics Reports. 2014;**540**(4):167-218

[16] Sevilla-Escoboza R et al. Error-feedback control of multistability. Journal of the Franklin Institute. 2017;**354**(16):7346-7358

[17] Magallón DA et al. Control of multistability in an erbium-doped fiber laser by an artificial neural network: A numerical approach. Mathematics. 2022;**10**(17):3140

[18] Echenausía-Monroy J et al. Multistability route in a PWL multi-scroll system through fractional-order derivatives. Chaos, Solitons & Fractals. 2022;**161**:112355

[19] Echenausía-Monroy J et al. Predicting the emergence of multistability in a Monoparametric PWL system. International Journal of Bifurcation and Chaos. 2022;**32**(14):2250206

[20] Brouwer LEJ. Über abbildung von mannigfaltigkeiten. Mathematische Annalen. 1911;**71**(1):97-115

[21] Poincare H. Surless courbes define barles equations differentiate less. Journal of Differential Equations. 1886;**2**:54-65

[22] Banach S. Sur les opérations dans les ensembles abstraits et leur application aux équations intégrales. Fundamenta Mathematicae. 1922;**3**(1):133-181

[23] Hunter JK, Nachtergaele B. Applied Analysis. Singapore: World Scientific Publishing Company; 2001

[24] Tarski A. A lattice-theoretical fixpoint theorem and its applications. Pacific Journal of Mathematics. 1955;**5**(2):285-309

[25] Wikipedia. Fixed Point (Mathematics). Available from: https://en.wikipedia.org/wiki/Fixed_point_(mathematics).

Section 2

Contractive and Nonexpansive Mappings

Chapter 2

Coupled Fixed Points for (φ, ψ)-Contractive Mappings in Partially Ordered Modular Spaces

Tayebe Lal Shateri

Abstract

The Banach contraction principle is the most famous fixed point theorem. Many authors presented some new results for contractions in partially ordered metric spaces. Fixed point theorems in modular spaces, generalizing the classical Banach fixed point theorem in metric spaces, have been studied extensively by many mathematicians. The aim of this paper is to determine some coupled fixed point theorems for nonlinear contractive mappings in the framework of a modular space endowed with a partial order. Our results are generalizations of the fixed point theorems due to M. Mursaleen, S.A. Mohiuddine and R.P. Agarwal.

Keywords: coupled fixed point, contraction, modular space, partially ordered modular space

1. Introduction

In 1922, Banach established the most famous fundamental fixed point theorem, so-called the Banach contraction principle [1], which has played an important role in various fields of applied mathematical analysis. Fixed point theory is one of the most important theory in mathematics. The Banach contraction mapping principle has many applications to very different type of problems arise in different branches. Many authors have obtained many interesting extensions and generalizations (cf. [2–8]).

The more generalization was given by Nakano [9] in 1950 based on replacing the particular integral form of the functional by an abstract one. This functional was called modular. In 1959, this idea, which was the basis of the theory of modular spaces and initiated by Nakano, was refined and generalized by Musielak and Orlicz [10]. Modular spaces have been studied for almost 40 years and there is a large set of known applications of them in various parts of analysis. For more details about modular spaces, we refer the reader to [11, 12].

Fixed point theorems in modular spaces, generalizing the classical Banach fixed point theorem in metric spaces, have been studied extensively by many mathematicians, see [13–18].

The author [19] has investigated some coupled coincidence and coupled common fixed point theorems for mixed g-monotone nonlinear contractive mappings in partially ordered modular spaces.

The aim of this paper is to determine some coupled fixed point theorems for (φ, ψ)- contractive mappings in the framework of partially ordered complete modular spaces. Our results are generalizations of the fixed point theorems due to M. Mursaleen, S.A. Mohiuddine and R.P. Agarwal [20]. First, we recall some basic definitions and notations about modular spaces from [11].

Definition 1.1. Let $\mathcal{X}$ be a vector space over $\mathbb{F}(=\mathbb{R}$ or $\mathbb{C})$.

A functional $\rho : \mathcal{X} \to [0, \infty]$ is said to be modular if for all $x,y \in \mathcal{X}$,

(i) $\rho(x) = 0$ if and only if $x = 0$,

(ii) $\rho(\alpha x) = \rho(x)$ for every $\alpha \in \mathbb{F}$ such that $|\alpha| = 1$,

(iii) $\rho(\alpha x + \beta y) \leq \rho(x) + \rho(y)$ if $\alpha,\beta \geq 0$ and $\alpha + \beta = 1$.

Definition 1.2. If in Definition 1.1, *(iii)* is replaced by

$$\rho(\alpha x + \beta y) \leq \alpha^s \rho(x) + \beta^s \rho(y), \tag{1}$$

for $\alpha, \beta \geq 0$, $\alpha + \beta = 1$ with an $s \in (0, 1]$, then we say that ρ is an s-convex modular, and if $s = 1$, ρ is said to be a convex modular.

Let ρ be a modular, we define the corresponding modular space, i.e. the vector space $\mathcal{X}_\rho$ given by

$$\mathcal{X}_\rho = \{x \in \mathcal{X} : \quad \rho(\lambda x) \to 0 \ \text{ as } \ \lambda \to 0\}. \tag{2}$$

The modular space $\mathcal{X}_\rho$ is a normed space with the Luxemburg norm, defined by

$$\|x\|_\rho = \inf\left\{\lambda > 0; \quad \rho\left(\frac{x}{\lambda}\right) \leq 1\right\}. \tag{3}$$

Definition 1.3. We say a function modular ρ satisfies the Δ_2–condition if there exists $\kappa > 0$ such that for any $x \in \mathcal{X}_\rho$, we have $\rho(2x) \leq \kappa\rho(x)$.

Definition 1.4. Let $\mathcal{X}_\rho$ be a modular space and suppose $\{x_n\}$ and x are in $\mathcal{X}_\rho$. Then.

i. $\{x_n\}$ is ρ–convergent to x and write $x_n \xrightarrow{\rho} x$ if $\rho(x_n - x) \to 0$ as $n \to \infty$.

ii. $\{x_n\}$ is ρ–Cauchy if $\rho(x_n - x_m) \to 0$ as $n,m \to \infty$.

iii. A subset $\mathcal{S}$ of $\mathcal{X}_\rho$ is called ρ–complete if any ρ–Cauchy sequence is ρ–convergent to an element of $\mathcal{S}$.

iv. The modular ρ has the Fatou property if $\rho(x) \leq \liminf_{n\to\infty}\rho(x_n)$ whenever $x_n \xrightarrow{\rho} x$.

Remark 1.5. *(ii)* A ρ-convergent sequence is ρ-cauchy if and only if ρ satisfies the Δ_2–condition. *(ii)* $\rho(.x)$ is an non-decreasing function, for any $x \in \mathcal{X}$. Fro this, let $0 < a < b$, putting $y = 0$ in *(iii)* of Definition 1.1 implies that

$$\rho(ax) = \rho\left(\frac{a}{b}bx\right) \leq \rho(bx),$$

for all $x \in \mathcal{X}$. Also, if ρ is a convex modular on $\mathcal{X}$ and $|\alpha| \leq 1$, then $\rho(\alpha x) \leq \alpha\rho(x)$ and $\rho(x) \leq \frac{1}{2}\rho(2x)$ for all $x \in \mathcal{X}$.

We end this section with a notion of a coupled fixed point introduced by Bhaskar and Lakshmikantham [5].

Definition 1.6. An element $(x, y) \in \mathcal{X} \times \mathcal{X}$ is called a coupled fixed point of the mapping $F : \mathcal{X} \times \mathcal{X} \to \mathcal{X}$ if

$$F(x, y) = x, \quad F(y, x) = y.$$

2. Coupled fixed point theorems for nonlinear (φ, ψ)-contractive type mappings

In this section, we establish some coupled fixed point results by considering (φ, ψ)-contractive mappings on modular spaces endowed with a partial order. We assume that ρ satisfies the Δ_2-condition with $\kappa < 1$.

Let Ψ' be the family of non-decreasing functions $\psi : [0, +\infty) \to [0, +\infty)$ such that

$$\sum_{n=1}^{\infty} \psi^n(t) < \infty, \ \psi^{-1}(\{0\}) = \{0\}, \ \psi(t) < t \ \text{ and } \lim_{r \to t^+} \psi(r) < t, \ forall t > 0. \tag{4}$$

The following results are generalizations of the fixed point theorems due to M. Mursaleen, S.A. Mohiuddine and R.P. Agarwal [20] in partially ordered modular spaces.

Definition 2.1. Let $(\mathcal{X}_\rho, \preceq)$ be a partially ordered modular space and $F : \mathcal{X} \times \mathcal{X} \to \mathcal{X}$ be a mapping. Then a map F is said to be (φ, ψ)-contractive if there exist two functions $\varphi : \mathcal{X}^2 \times \mathcal{X}^2 \to [0, \infty)$ and $\psi \in \Psi$ and there exist $\alpha, \beta > 0$ with $\alpha > \beta$ such that

$$\varphi((x, y), (z, w))\rho(\alpha(F(x, y) - F(z, w))) \leq \psi\left(\frac{\rho(\beta(x - z)) + \rho(\beta(y - w))}{2}\right) \tag{5}$$

for all $x, y, z, w \in \mathcal{X}$ with $x \succeq z$ and $y \preceq w$.

Definition 2.2. Let $F : \mathcal{X} \times \mathcal{X} \to \mathcal{X}$ and $\varphi : \mathcal{X}^2 \times \mathcal{X}^2 \to [0, \infty)$ be two mappings. Then F is called φ-admissible if

$$\varphi((x, y), (z, w)) \geq 1 \Rightarrow \varphi((F(x, y), F(y, x)), (F(z, w), F(w, z))) \geq 1 \tag{6}$$

for all $x, y, z, w \in \mathcal{X}$.

In the following theorem, we give some requirements that a φ-admissible mapping has a coupled fixed point.

Theorem 2.3. *Let $(\mathcal{X}, \preceq, \rho)$ be a complete ordered modular function space. Let $F : \mathcal{X} \times \mathcal{X} \to \mathcal{X}$ be a (φ, ψ)-contractive mapping having the mixed monotone property of $\mathcal{X}$. Suppose that.*

i. *F is φ-admissible,*

ii. *there exist $x_0, y_0 \in \mathcal{X}$ such that $x_0 \preceq F(x_0, y_0)$ and $y_0 \succeq F(y_0, x_0)$, also*

$$\varphi((x_0, y_0), (F(x_0, y_0), F(y_0, x_0))) \geq 1 \quad \text{and} \quad \varphi((y_0, x_0), (F(y_0, x_0), F(x_0, y_0))) \geq 1, \tag{7}$$

iii. *if $\{x_n\}$ and $\{y_n\}$ are sequences in $\mathcal{X}$ such that*

$$\varphi((x_n, y_n), (x_{n+1}, y_{n+1})) \geq 1 \quad and \quad \varphi((y_n, x_n), (y_{n+1}, x_{n+1})) \geq 1 \tag{8}$$

for all n and $\lim_{n\to\infty} x_n = x$ and $\lim_{n\to\infty} y_n = y$, then

$$\varphi((x_n, y_n), (x, y)) \geq 1 \quad and \quad \varphi((y_n, x_n), (y, x)) \geq 1. \tag{9}$$

Then F has a coupled fixed point.

Proof. Let $x_0, y_0 \in \mathcal{X}$ be such that

$$\varphi((x_0, y_0), (F(x_0, y_0), F(y_0, x_0))) \geq 1 \text{ and } \varphi((y_0, x_0), (F(y_0, x_0), F(x_0, y_0))) \geq 1 \tag{10}$$

and $x_0 \leq F(x_0, y_0)$ and $y_0 \geq F(y_0, x_0)$. Put $x_1 = F(x_0, y_0)$ and $y_1 = F(y_0, x_0)$. Let $x_2, y_2 \in \mathcal{X}$ be such that $x_2 = F(x_1, y_1)$ and $y_2 = F(y_1, x_1)$. Continuing this process, we can construct two sequences $\{x_n\}$ and $\{y_n\}$ in $\mathcal{X}$ such that

$$x_{n+1} = F(x_n, y_n) \quad \text{and} \quad y_{n+1} = F(y_n, x_n) \quad (n \geq 0). \tag{11}$$

Using the mathematical induction, we will show that

$$x_n \leq x_{n+1} \quad \text{and} \quad y_n \geq y_{n+1} \quad (n \geq 0). \tag{12}$$

By assumption, (12) hold for $n = 0$. Now suppose that (12) hold for some fixed $n \geq 0$. Then by the mixed monotone property of F, we have

$$x_{n+2} = F(x_{n+1}, y_{n+1}) \geq F(x_n, y_{n+1}) \geq F(x_n, y_n) = x_{n+1} \tag{13}$$

and

$$y_{n+2} = F(y_{n+1}, x_{n+1}) \leq F(y_n, x_{n+1}) \leq F(y_n, x_n) = y_{n+1}. \tag{14}$$

Hence (12) hold for $n \geq 0$. If for some n, $(x_{n+1}, y_{n+1}) = (x_n, y_n)$, then $F(x_n, y_n) = x_n$ and $F(y_n, x_n) = y_n$, and so F has a coupled fixed point. Thus we assumed that $(x_{n+1}, y_{n+1}) \neq (x_n, y_n)$ for all $n \geq 0$. Since F is φ-admissible, we have

$$\varphi((x_0, y_0), (x_1, y_1)) = \varphi(((x_0, y_0), (F(x_0, y_0), F(y_0, x_0))) \geq 1 \tag{15}$$

hence

$$\varphi((F(x_0, y_0), F(y_0, x_0)), (F(x_1, y_1), F(y_1, x_1))) = \varphi((x_1, y_1), (x_2, y_2)) \geq 1. \tag{16}$$

Therefore by induction we get

$$\varphi((x_n, y_n), (x_{n+1}, y_{n+1})) \geq 1 \quad \text{and} \quad \varphi((y_n, x_n), (y_{n+1}, x_{n+1})) \geq 1 \tag{17}$$

for all $n \in \mathbb{N}$. Since F is (φ, ψ)-contractive, using (35) and (17), we obtain

$$\begin{aligned} \rho(\alpha(x_n - x_{n+1})) &= \rho(\alpha(F(x_{n-1}, y_{n-1}), F(x_n, y_n))) \\ &\leq \varphi(((x_{n-1}, y_{n-1}), (x_n, y_n)))\rho(\alpha(F(x_{n-1}, y_{n-1}), F(x_n, y_n))) \\ &\leq \psi\left(\frac{\rho(\beta(x_{n-1} - x_n)) + \rho(\beta(y_{n-1} - y_n))}{2}\right), \end{aligned} \tag{18}$$

and

$$\begin{aligned}\rho(\alpha(y_n - y_{n+1})) &= \rho(\alpha(F(y_{n-1},\ x_{n-1}),\ F(y_n,\ x_n))) \\ &\le \varphi(((y_{n-1},\ x_{n-1}),\ (y_n,\ x_n)))\rho(\alpha(F(y_{n-1},\ x_{n-1}),\ F(y_n,\ x_n))) \\ &\le \psi\left(\frac{\rho(\beta(y_{n-1} - y_n)) + \rho(\beta(x_{n-1} - x_n))}{2}\right).\end{aligned} \tag{19}$$

Adding (18) and (23), we obtain

$$\frac{\rho(\alpha(x_n - x_{n+1})) + \rho(\alpha(y_n - y_{n+1}))}{2} \le \psi\left(\frac{\rho(\beta(x_{n-1} - x_n)) + \rho(\beta(y_{n-1} - y_n))}{2}\right) \tag{20}$$

Since $\beta < \alpha$ and ψ is non-decreasing, repeating the above process, we get

$$\frac{\rho(\alpha(x_n - x_{n+1})) + \rho(\alpha(y_n - y_{n+1}))}{2} \le \psi^n\left(\frac{\rho(\beta(x_0 - x_1)) + \rho(\beta(y_0 - y_1))}{2}\right) \tag{21}$$

for all $n \in \mathbb{N}$. Given $\varepsilon > 0$ there exists $N \in \mathbb{N}$ such that

$$\sum_{n \ge N} \psi^n\left(\frac{\rho(\beta(x_0 - x_1)) + \rho(\beta(y_0 - y_1))}{2}\right) < \frac{\varepsilon}{2}. \tag{22}$$

Let $m,n \in \mathbb{N}$ and $\alpha_0 \in \mathbb{R}^+$ be such that $m > n > N$ and $\frac{\beta}{\alpha} + \frac{1}{\alpha_0} = 1$. Then we have

$$\begin{aligned}\rho(\beta(x_n - x_m)) &\le \rho(\alpha(x_n - x_{n+1})) + \rho(\alpha_0\beta(x_{n+1} - x_m)) \\ &\le \rho(\alpha(x_n - x_{n+1})) + \kappa\rho(\beta(x_{n+1} - x_m)) \\ &\le \rho(\alpha(x_n - x_{n+1})) + \rho(\alpha(x_{n+1} - x_{n+2})) + (\alpha_0\beta(x_{n+2} - x_m)) \\ &\le \rho(\alpha(x_n - x_{n+1})) + \rho(\alpha(x_{n+1} - x_{n+2})) + (\beta(x_{n+2} - x_m)) \\ &\le \cdots \\ &\le \sum_{i=n}^{m-1} \rho(\alpha(x_i - x_{i+1})),\end{aligned} \tag{23}$$

similarly we obtain

$$\rho(\beta(y_n - y_m)) \le \sum_{i=n}^{m-1} \rho(\alpha(y_i - y_{i+1})). \tag{24}$$

Adding (23) and (24) we obtain

$$\begin{aligned}\frac{\rho(\beta(x_n - x_m)) + \rho(\beta(y_n - y_m))}{2} &\le \sum_{i=n}^{m-1} \frac{\rho(\alpha(x_i - x_{i+1})) + \rho(\alpha(y_i - y_{i+1}))}{2} \\ &\le \sum_{i=n}^{m-1} \psi^n\left(\frac{\rho(\beta(x_0 - x_1)) + \rho(\beta(y_0 - y_1))}{2}\right) \\ &< \frac{\varepsilon}{2}.\end{aligned} \tag{25}$$

Consequently

$$\rho(\beta(x_n - x_m)) \leq \rho(\beta(x_n - x_m)) + \rho(\beta(y_n - y_m)) < \varepsilon \tag{26}$$

and

$$\rho(\beta(y_n - y_m)) \leq \rho(\beta(x_n - x_m)) + \rho(\beta(y_n - y_m)) < \varepsilon, \tag{27}$$

therefore $\{x_n\}$ and $\{y_n\}$ are cauchy sequences in complete modular space $(\mathcal{X}, \rho)$, and so $\{x_n\}$ and $\{y_n\}$ are convergent in $(\mathcal{X}, \rho)$. Thus there exist $x, y \in \mathcal{X}$ such that

$$\lim_{n \to \infty} x_n = x \quad \text{and} \quad \lim_{n \to \infty} y_n = y.$$

Now from (17) and hypothesis *(iii)*, we get

$$\varphi((x_n, y_n), (x_{n+1}, y_{n+1})) \geq 1 \quad \text{and} \quad \varphi((y_n, x_n), (y_{n+1}, x_{n+1})) \geq 1 \tag{28}$$

for all $n \in \mathbb{N}$. From (28) and the condition *(iii)* of the modular ρ we obtain

$$\begin{aligned}
\rho(\beta(F(x, y) - x)) &\leq \rho(\alpha(F(x, y) - F(x_n, y_n))) + \rho(\alpha_0\beta(x_{n+1} - x)) \\
&\leq \varphi((x_n, y_n), (x, y))\rho(\alpha(F(x, y) - F(x_n, y_n))) + \rho(\alpha_0\beta(x_{n+1} - x)) \\
&\leq \psi\left(\frac{\rho(\beta(x_n - x)) + \rho(\beta(y_n - y))}{2}\right) + \rho(\alpha_0\beta(x_{n+1} - x)) \\
&< \frac{\rho(\beta(x_n - x)) + \rho(\beta(y_n - y))}{2} + \rho(\alpha_0\beta(x_{n+1} - x))
\end{aligned} \tag{29}$$

similarly, we get

$$\begin{aligned}
\rho(\beta(F(y, x) - y)) &\leq \rho(\alpha(F(y, x) - F(y_n, x_n))) + \rho(\alpha_0\beta(y_{n+1} - y)) \\
&\leq \varphi((y_n, x_n), (y, x))\rho(\alpha(F(y, x) - F(y_n, x_n))) + \rho(\alpha_0\beta(y_{n+1} - y)) \\
&\leq \psi\left(\frac{\rho(\beta(y_n - y)) + \rho(\beta(x_n - x))}{2}\right) + \rho(\alpha_0\beta(y_{n+1} - y)) \\
&< \frac{\rho(\beta(y_n - y)) + \rho(\beta(x_n - x))}{2} + \rho(\alpha_0\beta(y_{n+1} - y)).
\end{aligned} \tag{30}$$

Taking the limit as $n \to \infty$, we obtain

$$\rho(\beta(F(x, y) - x)) = 0 \quad \text{and} \quad \rho(\beta(F(y, x) - y)) = 0. \tag{31}$$

Therefore $F(x, y) = x$ and $F(y, x) = y$, that is F has a coupled fixed point. □

Remark 2.4. If in Theorem 2.3, we replace the property *(iii)* with the continuity of F, then the result holds that is F has a coupled fixed point. In fact, since F is continuous and $x_{n+1} = F(x_n, y_n)$ and $y_{n+1} = F(y_n, x_n)$, we get

$$x = \lim_{n \to \infty} x_n = \lim_{n \to \infty} F(x_{n-1}, y_{n-1}) = F(x, y) \tag{32}$$

and

$$y = \lim_{n\to\infty} y_n = \lim_{n\to\infty} F(y_{n-1}, x_{n-1}) = F(y, x). \tag{33}$$

Hence F has a coupled fixed point.

As the proof of Theorem 2.3, one can prove the following theorem.

Theorem 2.5. *In addition to the hypothesis of Theorem 2.3, suppose that for every $(x, y),(z, w)$ in $\mathcal{X} \times \mathcal{X}$, there exists (s, t) in $\mathcal{X} \times \mathcal{X}$ such that*

$$\varphi((x, y), (s, t)) \geq 1 \quad \textit{and} \quad \varphi((z, w), (s, t)) \geq 1, \tag{34}$$

and assume that (s, t) is comparable to (x, y) and (z, w). Then F has a unique coupled fixed point.

If we put $\psi(t) = mt$ for $m \in [0, 1)$ in Theorem 2.3, we obtain the following corollary.

Corollary 2.6. *Let $(\mathcal{X}, \leq, \rho)$ be a complete ordered modular function space. Let $F : \mathcal{X} \times \mathcal{X} \to \mathcal{X}$ be a (φ, ψ)-contractive mapping having the mixed monotone property of $\mathcal{X}$. Suppose that there exist $\alpha,\beta > 0$ with $\alpha > \beta$ such that*

$$\varphi((x, y), (z, w))\rho(\alpha(F(x, y) - F(z, w))) \leq \frac{m}{2}(\rho(\beta(x - z)) + \rho(\beta(y - w))) \tag{35}$$

for all $x, y, z, w \in \mathcal{X}$ with $x \geq z$ and $y \leq w$. Also if.

(ii) there exist $x_0, y_0 \in \mathcal{X}$ such that $x_0 \leq F(x_0, y_0)$ and $y_0 \geq F(y_0, x_0)$, also

$$\varphi((x_0, y_0), (F(x_0, y_0), F(y_0, x_0))) \geq 1 \quad \textit{and} \quad \varphi((y_0, x_0), (F(y_0, x_0), F(x_0, y_0))) \geq 1, \tag{36}$$

(ii) if $\{x_n\}$ and $\{y_n\}$ are sequences in $\mathcal{X}$ such that

$$\varphi((x_n, y_n), (x_{n+1}, y_{n+1})) \geq 1 \quad \textit{and} \quad \varphi((y_n, x_n), (y_{n+1}, x_{n+1})) \geq 1 \tag{37}$$

for all n and $\lim_{n\to\infty} x_n = x$ and $\lim_{n\to\infty} y_n = y$, then

$$\varphi((x_n, y_n), (x, y)) \geq 1 \quad \textit{and} \quad \varphi((y_n, x_n), (y, x)) \geq 1 \tag{38}$$

Then F has a coupled fixed point.

3. Conclusion

In the present paper, nonlinear contractive mappings in the framework of a modular space endowed with a partial order have been given, then some well-known coupled fixed point theorems in ordered metric spaces are extended to these mappings in modular spaces endowed with a partial order.

2010 Mathematics Subject Classification:

Primary 47H10, 54H25; Secondary 46A80

Author details

Tayebe Lal Shateri
Department of Mathematics and Computer Sciences, Hakim Sabzevari University, Sabzevar, Iran

*Address all correspondence to: t.shateri@hsu.ac.ir; shateri@ualberta.ca

References

[1] Banach S. Sur les opérations dans les ensembles abstraits et leur application aux équations intégrales. Fundamenta Mathematicae. 1922;**3**:133-181

[2] Abbas M, Nazir T, Radenović S. Fixed points of four maps in partially ordered metric spaces. Applied Mathematics Letters. 2011;**24**:1520-1526

[3] Agarwal RP, El-Gebeily MA, ÓRegan D. Generalized contractions in partially ordered metric spaces. Applicable Analysis. 2008;**87**:109-116

[4] Ahmad J, Arshad M, Vetro P. Coupled coincidence point results for ϕψ-contractive mappings in partially ordered metric spaces. Georgian Mathematical Journal. 2014;**21** (2):1-13

[5] Bhaskar TG, Lakshmikantham V. Fixed point theorems in partially ordered metric spaces and applications. Nonlinear Analysis. 2006;**65**:1379-1393

[6] Choudhury BS, Kundu A. A coupled coincidence point result in partially ordered metric spaces for compatible mappings. Nonlinear Analysis. 2010;**73**: 2524-2531

[7] Ćirić LB, Cakić N, Rajović M, Ume JS. Monotone generalized nonlinear contractions in partially ordered metric spaces. Fixed Point Theory and Applications. 2008;**2008**:131294

[8] Luong NV, Thuan NX. Coupled fixed points in partially ordered metric spaces and application. Nonlinear Analysis. 2011;**74**:983-992

[9] Nakano H. Modulared semi-ordered linear spaces. In: Tokyo Math. Book Ser. Vol. 1. Tokyo: Maruzen Co.; 1950

[10] Musielak J, Orlicz W. On Modular Spaces. Studia Mathematica. 1959;**18**:49-56

[11] Koslowski WM. Modular function spaces. New York, Basel: Dekker; 1988

[12] Musielak J. Orlicz Spaces and Modular Spaces, Lecture Notes in Math. Berlin: Springer; 1983. p. 1034

[13] Arandelović ID. On a fixed point theorem of Kirk. Journal of Mathematical Analysis and Applications. 2005;**301**(2):384-385

[14] Ćirić LB. A generalization of Banach's contraction principle. Proceedings of the American Mathematical Society. 1974;**45**(2):267-273

[15] Edelstein M. On fixed and periodic points under contractive mappings. Journal of the London Mathematical Society. 1962;**37**(1):74-79

[16] Kuaket K, Kumam P. Fixed point of asymptotic pointwise contractions in modular spaces. Applied Mathematics Letters. 2011;**24**:1795-1798

[17] Reich S. Fixed points of contractive functions. Bollettino dell'Unione Mathematica Italiana. 1972;**4**(5):26-42

[18] Wang X, Chen Y. Fixed points of asymptotic pointwise nonexpansive mappings in modular spaces. Applications of Mathematics. 2012; **2012**). Article ID 319394:6

[19] Shateri TL. Coupled fixed points theorems for non-linear contractions in partially ordered modular spaces. International Journal of Nonlinear

Analysis and Applications. 2020;**11**(2): 133-147

[20] Mursaleen M, Mohiuddine SA, Agarwal RP. Coupled fixed point theorems for $\alpha-\psi$-contractive type mappings in partially ordered metric spaces. Fixed Point Theory and Applications. 2012;**2012**. DOI: 10.1186/1687-1812-2012-228

Chapter 3

Common Fixed Points of Asymptotically Quasi-Nonexpansive Mappings in Cat(0) Spaces

Jamnian Nantadilok and Buraskorn Nuntadilok

Abstract

In this manuscript, we investigate and approximate common fixed points of two asymptotically quasi-nonexpansive mappings in CAT(0) spaces. Suppose $\mathbb{X}$ is a CAT (0) space and C is a nonempty closed convex subset of $\mathbb{X}$. Let $T_1, T_2 : C \to C$ be two asymptotically quasi-nonexpansive mappings, and $\mathbb{F} = F(T_1) \cap F(T_2) :=$ $\{x \in C : T_1 x = T_2 x = x\} \neq \varnothing$. Let $\{\alpha_n\}, \{\beta_n\}$ be sequences in [0,1]. If the sequence $\{x_n\}$ is generated iteratively by $x_{n+1} = (1-\alpha_n)x_n \oplus \alpha_n T_1^n y_n, y_n = (1-\beta_n)x_n \oplus \beta_n T_2^n x_n, n \geq 1$ and $x_1 \in C$ is the initial element of the sequence (A). We prove that $\{x_n\}$ converges strongly to a common fixed point of T_1 and T_2 if and only if $\lim_{n\to\infty} d(x_n, \mathbb{F}) = 0$. (B). Suppose $\{\alpha_n\}$ and $\{\beta_n\}$ are sequences in $[\varepsilon, 1-\varepsilon]$ forsome $\varepsilon \in (0, 1)$. If X is uniformly convex and if either T_2 or T_1 is compact, then $\{x_n\}$ converges strongly to some common fixed point of T_1 and T_2. Our results extend and improve the related results in the literature. We also give an example in support of our main results.

Keywords: asymptotically quasi-nonexpansive mappings, uniformly L-Lipschitzian mappings, fixed points, banach spaces, CAT(0) spaces

1. Introduction

Let C be a nonempty subset of a real normed linear space X. Let $T : C \to C$ be a self-mapping of C. Then T is said to be.

a. nonexpansive if $\|Tx - Ty\| \leq \|x - y\|$ for all $x,y \in C$;

b. quasi-nonexpansive if $F(T) \neq \varnothing$ and $\|Tx - p\| \leq \|x - p\|$ for all $x \in C$ and $p \in F(T)$ where $F(T) = \{x \in C : Tx = x\}$;

c. asymptotically nonexpansive with sequence $\{k_n\} \subset [0, \infty)$ if $\lim_{n\to\infty} k_n = 1$ and $\|T^n x - T^n y\| \leq k_n \|x - y\|$ for all $x,y \in C$ and $n \geq 1$;

d. symptotically quasi-nonexpansive with sequence $\{k_n\} \subset [0, \infty)$ if $F(T) \neq \emptyset$, $\lim_{n\to\infty} k_n = 1$ and $\|T^n x - p\| \leq k_n \|x - p\|$ for all $x \in C, p \in F(T)$ and $n \geq 1$.

It is clear that a nonexpansive mapping with $F(T) \neq \emptyset$ is quasi-nonexpansive and an asymptotically nonexpansive mapping with $F(T) \neq \emptyset$ is asymptotically quasi-nonexpansive. The converses are not true in general. The mapping T is said to be uniformly (L, γ)-Lipschitzian if there exists a constant $L > 0$ and $\gamma > 0$ such that $\|T^n x - T^n y\| \leq L\|x - y\|^\gamma$ for all $x, y \in C$ and $n \geq 1$.

The following example shows that there is a quasi-nonexpansive mapping which is not a nonexpansive mapping.

Example 1.1. (see [1]) Let $C = \mathbb{R}^1$ and define a mapping $T : C \to C$ by

$$Tx = \begin{cases} \dfrac{x}{2} & , \text{if } x \neq 0 \\ 0 & , \text{if } x = 0 \end{cases}$$

Then T is quasi-nonexpansive but not nonexpansive.

It is easy to see that a nonexpansive mapping is an asymptotically nonexpansive mapping with the sequence $\{k_n\} = \{1\}$.

It is easy to see that a quasi-nonexpansive mapping is an asymptotically quasi-nonexpansive mapping with the sequence $\{k_n\} = \{1\}$.

In 1972, Goebel and Kirk [2] introduced the class of asymptotically nonexpansive maps as a significant generalization of the class of nonexpansive maps. They proved that if the map $T : C \to C$ is asymptotically nonexpansive and C is a nonempty closed convex bounded subset of a uniformly convex Banach space X, then T has a fixed point. In [3], Goebel and Kirk extended this result to the broader class of uniformly $(L, 1)$-Lipschitzian mappings with $L < \lambda$ and, where λ is sufficiently near 1 (but greater than 1).

Iterative approximation of fixed points of nonexpansive mappings and their generalizations (asymptotically nonexpansive mappings, etc.) have been investigated by a number of authors (see, [4–21] for examples) via the Mann iterates or the Ishikawa-type iteration.

Later, in 2001 Khan and Takahashi [22] studied the problem of approximating common fixed points of two asymptotically nonexpansive mappings. In 2002, Qihou [23] also established a strong convergence theorem for the Ishikawa-type iterative sequences with errors for a uniformly (L, γ)-Lipschitzian asymptotically nonexpansive self-mapping of a nonempty compact convex subset of a uniformly convex Banach space.

Recently, in 2005 Shahzad and Udomene [24] investigated the approximation of common fixed points of two asymptotically quasi-nonexpansive mappings in Banach spaces. More precisely, they obtained the following results.

Theorem 1.2. [24] *Let C be a nonempty closed convex subset of a real Banach space X. Let $T_1, T_2 : C \to C$ be two asymptotically quasi-nonexpansive mappings with sequences $\{u_n\}, \{v_n\} \subset [0, \infty)$ such that $\sum_{n=1}^{\infty} u_n < \infty$ and $\sum_{n=1}^{\infty} v_n < \infty$, and $\mathbb{F} = F(T_1) \cap F(T_2) := \{x \in C : T_1 x = T_2 x = x\} \neq \emptyset$. Let $x_1 \in C$ be arbitrary, define the sequence $\{x_n\}$ iteratively by the iteration*

$$\begin{aligned} x_{n+1} &= (1 - \alpha_n) x_n + \alpha_n T_1^n y_n \\ y_n &= (1 - \beta_n) x_n + \beta_n T_2^n x_n, \end{aligned} \tag{1}$$

for all $n \geq 1$, where $\{\alpha_n\}$ and $\{\beta_n\}$ are sequences in $[0, 1]$. Then.

1. $\|x_{n+1} - p\| \leq (1 - \gamma_n)\|x - p\|$ for all $n \geq 1, p \in \mathbb{F}$, and for some sequence $\{\gamma_n\}$ of numbers with $\sum_{n=1}^{\infty} \gamma_n < \infty$.

2. There exists a constant $K > 0$ such that $\|x_{n+m} - p\| \leq K\|x_n - p\|$ for all $n,m \geq 1$ and $p \in \mathbb{F}$.

Theorem 1.3. [24] *Let C be a nonempty closed convex subset of a real Banach space X. Let $T_1, T_2 : C \to C$ be two asymptotically quasi-nonexpansive mappings with sequences $\{u_n\},\{v_n\} \subset [0, \infty)$ such that $\sum_{n=1}^{\infty} u_n < \infty$ and $\sum_{n=1}^{\infty} v_n < \infty$, and $\mathbb{F} = F(T_1) \cap F(T_2) \neq \emptyset$. Let $\{\alpha_n\},\{\beta_n\} \subset [0, 1]$. Define the sequence $\{x_n\}$ as in (1) and $x_1 \in C$ is the initial element of the sequence. Then $\{x_n\}$ converges strongly to a common fixed point of T_1 and $T_2 \Leftrightarrow \lim \ \inf_{n \to \infty} d(x_n, \mathbb{F}) = 0$..*

Theorem 1.4. [24] *Let X be a real uniformly convex Banach space and C a nonempty closed convex subset of X. Let $T_1, T_2 : C \to C$ be two uniformly continuous asymptotically quasi-nonexpansive mappings with sequences $\{u_n\},\{v_n\} \subset [0, \infty)$ such that $\sum_{n=1}^{\infty} u_n < \infty$, $\sum_{n=1}^{\infty} v_n < \infty$, and $\mathbb{F} = F(T_1) \cap F(T_2) \neq \emptyset$. Let $\{\alpha_n\}$ and $\{\beta_n\}$ be sequences in $[\varepsilon, 1 - \varepsilon]$ forsome $\varepsilon \in (0, 1)$. Define the sequence $\{x_n\}$ as in* (1) *and $x_1 \in C$ is the initial element of the sequence. Assume, in addition, that either T_2 or T_1 is compact. Then $\{x_n\}$ converges strongly to a common fixed point of T_1 and T_2.*

2. Preliminaries

In this section, we present some basic facts about the CAT(0) spaces and hyperbolic spaces with some useful results which are required in the sequel. The connection between CAT(0) spaces and hyperbolic spaces presented here would help, at least for beginners, to appreciate the main results presented in this manuscript.

2.1 CAT(0) spaces

Let (X, d) be a metric space. A geodesic path joining $x \in X$ to $y \in X$ (or, more briefly, a geodesic from x to y) is a map $\omega : [0, a] \to X, [0, a] \subset R$ such that $\omega(0) = x, \omega(a) = y$, and $d(\omega(m), \omega(n)) = |m - n|$ for all $m,n \in [0, a]$. In particular, ω is an isometry and $d(x, y) = a$. The image α of ω is called a geodesic (or metric) segment joining x and y. A unique geodesic segment from x to y is denoted by $[x, y]$. The space (X, d) is called to be a geodesic space if every two points of X are joined by a geodesic, and X is said to be uniquely geodesic if there is exactly one geodesic joining x and y for each $x,y \in X$. If $Y \subseteq X$ then Y is said to be convex if Y includes every geodesic segment joining any two of its points. If (X, d) is a geodesic metric space, a *geodesic triangle* $\Delta(a_1, a_2, a_3)$ consists of three points a_1,a_2,a_3 in X (the vertices of Δ) and a geodesic segment between each pair of vertices (the edges of Δ). A comparison triangle for geodesic triangle $\Delta(a_1, a_2, a_3)$ in (X, d) is a triangle $\overline{\Delta}(a_1, a_2, a_3) := \Delta(\overline{a}_1, \overline{a}_2, \overline{a}_3)$ in the Euclidean plane $\mathbb{R}^2$ satisfying $d_{\mathbb{R}^2}(\overline{a}_i, \overline{a}_j) = d(a_i, a_j)$ for $i,j \in 1,2,3$. Such a triangle always exists (See [25]).

Definition 2.1. A geodesic space (X, d) is said to be a CAT(0) space if for any geodesic triangle $\Delta \subset X$ and $a,b \in \Delta$ we have $d(a, b) \leq d(\overline{a}, \overline{b})$ where $\overline{a},\overline{b} \in \overline{\Delta}$.

Remark 2.2. Any complete, simply connected Riemannian manifold having nonpositive sectional curvature is a CAT(0) space. Other examples of CAT(0) spaces include pre-Hilbert spaces, R-trees, Euclidean buildings, and the complex Hilbert ball with a hyperbolic metric, (see [25–27] for example).

Definition 2.3. A geodesic triangle $\Delta(p, q, r)$ in (X, d) is said to satisfy the $CAT(0)$ *inequality* if for any $u,v \in \Delta(p, q, r)$ and for their comparison points $\overline{u},\overline{v} \in \overline{\Delta}(\overline{p}, \overline{q}, \overline{r})$, one has

$$d(u, v) \leq d_{\mathbb{R}^2}(\overline{u}, \overline{v}).$$

For other equivalent definitions and basic properties of $CAT(0)$ spaces, we refer the readers to standard texts, such as ref. [25].

Note that if x,a_1,a_2 are points of $CAT(0)$ space and if a_0 is the midpoint of the segment $[a_1, a_2]$ (we write $a_0 = \frac{1}{2}a_1 \oplus \frac{1}{2}a_2$), then the $CAT(0)$ inequality implies

$$d(x, a_0)^2 = d\left(x, \frac{1}{2}a_1 \oplus \frac{1}{2}a_2\right) \leq \frac{1}{2}d(x, a_1)^2 + \frac{1}{2}d(x, a_2)^2 - \frac{1}{4}d(a_1, a_2)^2 \tag{2}$$

The inequality (2) is called the **CN inequality of Bruhat and Tits** [28]. We refer readers to some brilliant known CAT(0) space results in [29–33] and references therein.

We now collect some useful facts about CAT(0) spaces, which will be used frequently in the proof of our main results.

Lemma 2.4. *(See* [31]*) Let (X, d) be a CAT(0) space.*

i. *For $x_1,x_2 \in X$ and $\alpha \in [0, 1]$, there exists a unique point $y \in [x_1, x_2]$ such that*

$$d(x_1, y) = \alpha d(x_1, x_2) \text{ and } d(x_2, y) = (1 - \alpha)d(x_1, x_2). \tag{3}$$

We write $y = (1 - \alpha)x_1 \oplus \alpha x_2$ for the unique point y satisfying (3).

ii. *For $x,y,z \in X$ and $\alpha \in [0, 1]$, we have*

$$d((1 - \alpha)x \oplus \alpha y, z) \leq (1 - \alpha)d(x, z) + \alpha d(y, z).$$

iii. *For $x,y,z \in X$ and $\alpha \in [0, 1]$ we have*

$$d((1 - \alpha)x \oplus \alpha y, z)^2 \leq (1 - \alpha)d(x, z)^2 + \alpha d(y, z)^2 - \alpha(1 - \alpha)d(x, y)^2.$$

Lemma 2.5. *(See* [34]*) Let $\{\alpha_n\}$,$\{\beta_n\}$ be two sequences such that.*

i. $0 \leq \alpha_n,\beta_n < 1$,

ii. $\beta_n \to 0$ *and* $\sum \alpha_n\beta_n = \infty$.

Let $\{\gamma_n\}$ be a nonnegative real sequence such that $\sum \alpha_n\beta_n(1 - \beta_n)\gamma_n$ is bounded. Then $\{\gamma_n\}$ has a subsequence that converges to zero.

Lemma 2.6. *(see,* [17]*). Let $\{\lambda_n\}$ and $\{\sigma_n\}$ be sequences of nonnegative real numbers such that $\lambda_{n+1} \leq \lambda_n + \sigma_n$, $\forall\, n \geq 1$ and $\sum_{n=1}^{\infty}\sigma_n < \infty$. Then $\lim_{n\to\infty} \lambda_n$ exists. Moreover, if there exists a subsequence $\left\{\lambda_{n_j}\right\}$ of $\{\lambda_n\}$ such that $\lambda_{n_j} \to 0$ as $j \to \infty$, then $\lambda_n \to 0$ as $n \to \infty$..*

2.2 Hyperbolic spaces

In this section, we recall some notions of hyperbolic spaces. This class of spaces contains the class of CAT(0) spaces (See [35, 36]).

Definition 2.7. (See [36]) Let (X, d) be a metric space and $\mathcal{W} : X \times X \times [0, 1] \to X$ be a mapping satisfying:-.

$\mathcal{W}1.\ d(z, \mathcal{W}(x,\ y,\ \alpha)) \le (1-\alpha)d(z, x) + \alpha d(z, y)$,.

$\mathcal{W}2.\ d(\mathcal{W}(x,\ y,\ \alpha), \mathcal{W}(x,\ y,\ \beta)) = |\alpha - \beta| d(x, y)$,

$\mathcal{W}3.\ \mathcal{W}(x, y, \alpha) = \mathcal{W}(y, x, (1-\alpha))$,.

$\mathcal{W}4.\ d(\mathcal{W}(x,\ z,\ \alpha), \mathcal{W}(y,\ w,\ \alpha)) \le (1-\alpha)d(x, y) + \alpha d(z, w)$.

for all $x,y,z,w \in X, \alpha, \beta \in [0, 1]$. We call the triple $(X, d, \mathcal{W})$ a **hyperbolic space**.

It follows from $(\mathcal{W}1.)$ that, for each $x,y \in X$ and $\alpha \in [0, 1]$,

$$d(x, \mathcal{W}(x,\ y,\ \alpha)) \le \alpha d(x, y), d(y, \mathcal{W}(x,\ y,\ \alpha)) \le (1-\alpha)d(x, y) \tag{4}$$

In fact, we can get that (see [33]),

$$d(x, \mathcal{W}(x,\ y,\ \alpha)) = \alpha d(x, y), d(y, \mathcal{W}(x,\ y,\ \alpha)) = (1-\alpha)d(x, y). \tag{5}$$

Similar to (3), we can also use the notation $(1-\alpha)x \oplus \alpha y$ for such a point $\mathcal{W}(x, y, \alpha)$ in **hyperbolic space**.

A mapping $\eta : (0, \infty) \times (0, 2] \to (0, 1]$ providing such a $\delta := \eta(r, \varepsilon)$ forgiven $r > 0$ and $\varepsilon \in (0, 2]$ is called a **modulus of uniform convexity**.

Definition 2.8. (See [37, 38]) Let (X, d) be a hyperbolic metric space. X is said to be uniformly convex whenever $\delta(r, \varepsilon) > 0$, for any $r > 0$ and $\varepsilon > 0$, where

$$\delta(r, \varepsilon) = \inf \left\{ 1 - \frac{1}{r} d\left(\frac{1}{2}x \oplus \frac{1}{2}y,\ a \right) : d(x, a) \le r, d(y, a) \le r, d(x, y) \ge r\varepsilon \right\}$$

for any $a \in X$.

Note that if X is a uniformly convex hyperbolic space, then for every $s \ge 0$ and $\varepsilon > 0$, there exists $\eta(s, \varepsilon) > 0$ such that $\delta(r, \varepsilon) > \eta(s, \varepsilon) > 0$ for any $r > s$. One can see that $\delta(r, 0) = 0$. Moreover $\delta(r, \varepsilon)$ is an increasing function of ε.

The following result is very useful which is an analog of Shu ([15], Lemma 1.3). It can be applied to a CAT(0) space as well.

Lemma 2.9. *(See [33, 39]) Let (X, d) be a uniformly convex hyperbolic space. Let $\{x_n\}, \{y_n\}$ be sequences in X and $c \in [0, +\infty)$ be such that* $\limsup_{n\to\infty} d(x_n, a) \le c, \limsup_{n\to\infty} d(y_n, a) \le c$, *and* $\lim_{n\to\infty} d((1-\alpha_n)x_n \oplus \alpha_n y_n, a) = c$, *where* $\alpha_n \in [a, b]$, *with* $0 < a \le b < 1$. *Then* $\lim_{n\to\infty} d(x_n, y_n) = 0$..

Inspired and motivated by Shahzad and Udomene [24], the purpose of this paper is to establish common fixed point theorems for two asymptotically quasi-nonexpansive mappings in the setting of CAT(0) spaces. Our results significantly extend and improve the results obtained by Shahzad and Udomene in ref. [24], as well as the related results in the existing literature.

3. Main results

In this section, we let X denote a CAT(0) space and C be a nonempty closed convex subset of a CAT(0) space X. Let $T_1, T_2 : C \to C$ be two asymptotically quasi-nonexpansive mappings with sequences $\left\{k_n^{(i)}\right\} \subset [1, \infty)$ satisfying $\sum_{n=1}^{\infty}\left(k_n^{(i)} - 1\right) < \infty, (i = 1, 2)$, respectively. Put $k_n = \max\left\{k_n^{(1)}, k_n^{(2)}\right\}$, then obviously $\sum_{n=1}^{\infty}(k_n - 1) < \infty$. From now on we will take this sequence $\{k_n\}$ for both T_1 and T_2. Recall that $F(T) = \{x : Tx = x\}$ and $\mathbb{F} := F(T_1) \cap F(T_2) = \{x \in C : T_1x = T_2x = x\}$. Following ref. [24], we introduce the following iterative scheme in the setting of CAT (0) space. Starting from arbitrary $x_1 \in C$,

$$\begin{aligned} x_{n+1} &= (1 - \alpha_n)x_n \oplus \alpha_n T_1^n y_n \\ y_n &= (1 - \beta_n)x_n \oplus \beta_n T_2^n x_n, \end{aligned} \tag{6}$$

for all $n \geq 1$, where $\{\alpha_n\}$ and $\{\beta_n\}$ are sequences in $[0, 1]$.

Lemma 3.1. *Let $(\mathbb{X}, d)$ be a CAT(0) space and C a nonempty closed convex subset of* $\mathbb{X}$. *Let $T_1, T_2 : C \to C$ be two asymptotically quasi-nonexpansive mappings and $\mathbb{F} = F(T_1) \cap F(T_2) \neq \emptyset$. Let $\{\alpha_n\}$ and $\{\beta_n\}$ be sequences in* $[0,1]$. *Define the sequence $\{x_n\}$ by iteration* (6). *Then.*

i. $d(x_{n+1}, p) \leq (1 + \gamma_n)d(x_n, p)$ for all $n \geq 1, p \in \mathbb{F}$, for some sequence of numbers $\{\gamma_n\}$ with $\sum_{n=1}^{\infty}\gamma_n < \infty$.

ii. there exists a constant $K > 0$ such that $d(x_{n+m}, p) \leq Kd(x_n, p)$ for all $n, m \geq 1$ and $p \in \mathbb{F}$.

Proof: (i). Taking $p \in \mathbb{F}$. Let $y_n = (1 - \beta_n)x_n \oplus \beta_n T_2^n x_n$. From (6) and by using Lemma 2.4(ii) we get

$$\begin{aligned} d(x_{n+1}, p) &= d\left((1 - \alpha_n)x_n \oplus \alpha_n T_1^n y_n, p\right) \\ &\leq (1 - \alpha_n)d(x_n, p) + \alpha_n d\left(T_1^n y_n, p\right) \\ &\leq (1 - \alpha_n)d(x_n, p) + \alpha_n k_n d\left(y_n, p\right) \\ &= (1 - \alpha_n)d(x_n, p) + \alpha_n k_n d\left((1 - \beta_n)x_n \oplus \beta_n T_2^n x_n, p\right) \\ &\leq (1 - \alpha_n)d(x_n, p) + \alpha_n k_n\left[(1 - \beta_n)d(x_n, p) + \beta_n d\left(T_2^n x_n, p\right)\right] \\ &\leq (1 - \alpha_n)d(x_n, p) + \alpha_n k_n\left[(1 - \beta_n)d(x_n, p) + \beta_n k_n d(x_n, p)\right] \\ &= \left((1 - \alpha_n + \alpha_n k_n - \alpha_n k_n \beta_n + \alpha_n \beta_n k_n^2\right)d(x_n, p) \\ &\leq \left((1 + \alpha_n k_n + \alpha_n \beta_n k_n^2\right)d(x_n, p) \\ &= (1 + \gamma_n)d(x_n, p) \end{aligned} \tag{7}$$

where $\gamma_n = \alpha_n k_n + \alpha_n \beta_n k_n^2$ with $\sum_{n=1}^{\infty} b_n < \infty$.

i. We know that $1 + x \leq \exp(x)$, for all $x \geq 0$. Notice that for any $n, m \geq 1$,

$$\begin{aligned} d(x_{n+m}, p) &\leq (1 + b_{n+m-1})d(x_{n+m-1}, p) \\ &\leq \exp(b_{n+m-1})d\left(x_{n+m-1,p}\right) \\ &\leq \exp(b_{n+m-1} + b_{n+m-2})d\left(x_{n+m-2,p}\right) \\ &\vdots \\ &\leq \exp\left(\sum_{k=n}^{n+m-1} b_k\right)d(x_n, p). \end{aligned} \tag{8}$$

Taking $K = \exp\left(\sum_{k=n}^{\infty} b_k\right)$. Then $0 < K < \infty$, we obtain

$$d(x_{n+m}, p) \leq K d(x_n, p) \tag{9}$$

where $p \in \mathbb{F}$. This completes our proof.

Theorem 3.2. *Let* $(\mathbb{X}, d)$ *be a complete* $CAT(0)$ *space and C a nonempty closed convex subset of X. Let* $T_1, T_2 : C \to C$ *be two asymptotically quasi-nonexpansive mappings* (T_1 *and* T_2 *need not be continuous), and* $\mathbb{F} = F(T_1) \cap F(T_2) \neq \emptyset$. *Let* $\{\alpha_n\}, \{\beta_n\}$ *be sequences in* $[0,1]$. *From arbitrary* $x_1 \in C$, *define the sequence* $\{x_n\}$ *by iteration* (6). *Then* $\{x_n\}$ *converges strongly to a common fixed point of* T_1 *and* T_2 *if and only if* $\lim_{n\to\infty} d(x_n, \mathbb{F}) = 0$.

Proof: The necessary conditions are obvious. We shall only prove the sufficient condition. By Lemma 3.1, we have $d(x_{n+1}, p) \leq (1 + \gamma_n) d(x_n, p)$ for all $n \geq 1$ and $p \in \mathbb{F}$. Therefore,

$$d(x_{n+1}, \mathbb{F}) \leq (1 + \gamma_n) d(x_n, \mathbb{F}).$$

Since $\sum_{n=1}^{\infty} \gamma_n < \infty$ and $\lim \inf_{n\to\infty} d(x_n, \mathbb{F}) = 0$, from Lemma 2.6 we deduce that $\lim_{n\to\infty} d(x_n, \mathbb{F}) = 0$. Next, we show that the sequence $\{x_n\}$ is Cauchy. Since $\lim_{n\to\infty} d(x_n, \mathbb{F}) = 0$, given any $\varepsilon > 0$, there exists a positive number N_0 such that $d(x_n, \mathbb{F}) < \frac{\varepsilon}{4K}$ for all $n \geq N_0$, where $K > 0$ is the constant in Lemma 3.1(2). So we can find $q \in \mathbb{F}$ such that $d(x_{N_0}, q) \leq \frac{\varepsilon}{3K}$. Again by Lemma 3.1(2), we have that

$$\begin{aligned} d(x_{n+m}, x_n) &\leq d(x_{n+m}, q) + d(x_n, q) \\ &\leq K d(x_{N_0}, q) + K d(x_{N_0}, q) \\ &= 2K d(x_{N_0}, q) < \varepsilon. \end{aligned} \tag{10}$$

for all $n \geq N_0$ and $m \geq 1$. This implies that $\{x_n\}$ is Cauchy and so is convergent since $\mathbb{X}$ is complete. Hence, $\{x_n\}$ is a Cauchy sequence in a closed convex subset C of a CAT(0) space $\mathbb{X}$, therefore, it must converge to a point in C. Let $\lim_{n\to\infty} x_n = q'$.
Now, $\lim_{n\to\infty} d(x_n, \mathbb{F}) = 0$ yields that $d(q', \mathbb{F}) = 0$. Since the set of fixed points of asymptotically nonexpansive mappings is closed, we have $q' \in \mathbb{F}$. This completes our proof.

Lemma 3.3. *Let* $(\mathbb{X}, d)$ *be a* $CAT(0)$ *space and C a nonempty closed convex subset of X. Let* $T_1, T_2 : C \to C$ *be two uniformly continuous asymptotically quasi-nonexpansive mappings, and* $\mathbb{F} = F(T_1) \cap F(T_2) \neq \emptyset$. *Let* $\{\alpha_n\}$ *and* $\{\beta_n\}$ *be sequences in* $[\varepsilon, 1 - \varepsilon]$ *forsome* $\varepsilon \in (0, 1)$. *From arbitrary* $x_1 \in C$, *define the sequence* $\{x_n\}$ *by iteration* (6). *Then*

$$\lim_{n\to\infty} d(x_n, T_2^n x_n) = \lim_{n\to\infty} d(x_n, T_1^n x_n) = \lim_{n\to\infty} d(x_n, T_1^n y_n) = 0. \tag{11}$$

Proof: Let $p \in \mathbb{F}$. Then, by Lemma 3.1(1) and Lemma 2.6 $\lim_{n\to\infty} d(x_n, p)$ exists. Suppose $\lim_{n\to\infty} d(x_n, p) = r$. If $r = 0$, then by the continuity of T_1 and T_2 the conclusion follows. Now suppose $r > 0$. We claim

$$\lim_{n\to\infty} d(x_n, T_1^n y_n) = \lim_{n\to\infty} d(x_n, T_1^n x_n) = \lim_{n\to\infty} d(x_n, T_2^n x_n) = 0. \tag{12}$$

From $y_n = (1-\beta_n)x_n \oplus \beta_n T_2^n x_n$. Since $\{x_n\}$ is bounded, there exists $R>0$ such that $x_n - p, y_n - p \in B_R(0)$ for all $n \geq 1$. Using Lemma 2.4(iii), we have that

$$\begin{aligned} d(y_n, p)^2 &= d((1-\beta_n)x_n \oplus \beta_n T_2^n x_n, p)^2 \\ &\leq (1-\beta_n)d(x_n,p)^2 + \beta_n d(T_2^n x_n, p)^2 - \beta_n(1-\beta_n)d(x_n, T_2^n x_n)^2 \\ &\leq (1-\beta_n)d(x_n,p)^2 + \beta_n k_n^2 d(x_n,p)^2 - \beta_n(1-\beta_n)d(x_n, T_2^n x_n)^2 \\ &\leq (1+\beta_n(k^2-1))d(x_n,p)^2 \leq d(x_n,p)^2. \end{aligned} \tag{13}$$

Again by Lemma 2.4(iii), it follows that

$$\begin{aligned} d(x_{n+1}, p)^2 &= d((1-\alpha_n)x_n \oplus \alpha_n T_1^n y_n, p)^2 \\ &\leq (1-\alpha_n)d(x_n,p)^2 + \alpha_n d(T_1^n x_n, p)^2 - \alpha_n(1-\alpha_n)d(x_n, T_1^n y_n)^2 \\ &\leq (1-\alpha_n)d(x_n,p)^2 + \alpha_n k_n^2 d(x_n,p)^2 - \alpha_n(1-\alpha_n)d(x_n, T_1^n y_n)^2. \end{aligned} \tag{14}$$

Equivalently

$$\begin{aligned} \alpha_n(1-\alpha_n)d(x_n, T_1^n y_n)^2 &\leq (1+\alpha_n(k_n^2-1))d(x_n,p)^2 - d(x_{n+1},p)^2 \\ &\leq (1+(k_n^2-1))d(x_n,p)^2 - d(x_{n+1},p)^2 \\ &= d(x_n,p)^2 - d(x_{n+1},p)^2. \end{aligned} \tag{15}$$

Summing up the first m term of the above inequality, we get

$$\sum_{n=1}^{m} \alpha_n(1-\alpha_n)d(x_n, T_1^n y_n)^2 \leq d(x_1,p)^2 - d(x_{m+1},p)^2 < \infty \tag{16}$$

for all $m \geq 1$. Now (16) implies that

$$\sum_{n=1}^{\infty} \alpha_n(1-\alpha_n)d(x_n, T_1^n y_n)^2 < \infty. \tag{17}$$

Since $0 \leq \alpha_n(1-\alpha_n) < 1$, $d(x_n, T_1^n x_n)^2 \to 0$ as $n \to \infty$. Therefore, we obtain

$$\lim_{n\to\infty} d(x_n, T_1^n y_n) = 0. \quad (*) \tag{18}$$

Since T_1 is asymptotically quasi-nonexpansive, we can get that $d(T_1^n y_n, p)) \leq k_n d(y_n, p)$ for all $n \in \mathbb{N}$. From (13), we have that

$$\limsup_{n\to\infty} d(T_1^n y_n, p) \leq r. \tag{19}$$

Similarly, we get

$$\limsup_{n\to\infty} d(T_2^n x_n, p) \leq r. \tag{20}$$

One can see that

$$\limsup_{n\to\infty} d(x_n, p) \le \lim_{n\to\infty} d(x_n, p) = r. \tag{21}$$

Since T_1 is asymptotically quasi-nonexpansive, we get

$$\begin{aligned} d(x_n, p) &\le d\left(x_n, T_1^n y_n\right) + d\left(T_1^n y_n, p\right) \\ &\le d\left(x_n, T_1^n y_n\right) + k_n d\left(y_n, p\right). \end{aligned} \tag{22}$$

Taking the limit inferior to above inequality and from (18), we obtain

$$r \le \liminf_{n\to\infty} d\left(y_n, p\right). \tag{23}$$

On the other hand, by Lemma 2.4(ii) we have

$$\begin{aligned} d\left(y_n, p\right) &= d\left((1-\beta_n)x_n \oplus \beta_n T_2^n x_n, p\right) \\ &\le (1-\beta_n) d(x_n, p) + \beta_n d\left(T_2^n x_n, p\right) \\ &= [(1 + \beta_n(k_n - 1)] d(x_n, p) \end{aligned} \tag{24}$$

which implies

$$\limsup_{n\to\infty} d\left(y_n, p\right) \le r. \tag{25}$$

This gives

$$\lim_{n\to\infty} d\left((1-\alpha_n)x_n \oplus \alpha_n T_2^n x_n, p\right) = r. \tag{26}$$

Using (20), (21), (26), and Lemma 2.9, we obtain

$$\lim_{n\to\infty} d\left(x_n, T_2^n x_n\right) = 0. \ (*) \tag{27}$$

From (23) and (25), we obtain

$$\lim_{n\to\infty} d\left(y_n, p\right) = r. \tag{28}$$

On the other hand, consider

$$\begin{aligned} d(x_{n+1}, p) &= d\left((1-\alpha_n)x_n \oplus \alpha_n T_1^n y_n, p\right) \\ &\le (1-\alpha_n) d(x_n, p) + \alpha_n k_n d\left(y_n, p\right). \end{aligned} \tag{29}$$

This implies

$$\lim_{n\to\infty} d\left((1-\alpha_n)x_n \oplus \alpha_n T_1^n y_n, p\right) = r. \tag{30}$$

From (19), (21), (30), and by Lemma 2.9, we also obtain

$$\lim_{n\to\infty} d\left(x_n, T_1^n y_n\right) = 0. \tag{31}$$

Next, we show $\lim_{n\to\infty} d\left(x_n, T_1^n x_n\right) = 0.$.

Consider

$$d(x_n, y_n) \le d(x_n, (1-\beta_n)x_n \oplus \beta_n T_2^n x_n) \le (1-\beta_n)d(x_n, x_n) + \beta_n d(x_n, T_2^n x_n) \to 0 \text{ as } n \to \infty \tag{32}$$

and

$$d(T_1^n x_n, x_n) \le d(T_1^n x_n, T_1^n y_n) + d(T_1^n y_n, x_n). \tag{33}$$

Since T_1 is uniformly continuous and $d(x_n, y_n) \to 0$ as $n \to \infty$, it follows from (32) and (33) that

$$\lim_{n\to\infty} d(T_1^n x_n, x_n) = 0. \ (*)$$

Our proof is finished.

Theorem 3.4. *Let $(\mathbb{X}, d)$ be a CAT(0) space and C a nonempty closed convex subset of $\mathbb{X}$. Let $T_1, T_2 : C \to C$ be two uniformly continuous asymptotically quasi-nonexpansive mappings, and $\mathbb{F} = F(T_1) \cap F(T_2) \neq \emptyset$. Let $\{\alpha_n\}$ and $\{\beta_n\}$ be sequences in $[\varepsilon, 1-\varepsilon]$ forsome $\varepsilon \in (0, 1)$. From arbitrary $x_1 \in C$, define the sequence $\{x_n\}$ by the recursion (6). Assume, in addition, that either T_2 or T_1 is compact. Then $\{x_n\}$ converges strongly to some common fixed point of T_1 and T_2.*

Proof: By Lemma 3.3, we have

$$\lim_{n\to\infty} d(x_n, T_1^n x_n) = 0 = \lim_{n\to\infty} d(x_n, T_2^n x_n) \tag{34}$$

and also

$$\lim_{n\to\infty} d(x_n, T_1^n y_n) = 0. \tag{35}$$

If T_2 is compact, then there exists a subsequence $\{T_2^{n_k} x_{n_k}\}$ of $\{T_2^n x_n\}$ such that $T_2^{n_k} x_{n_k} \to p$ as $k \to \infty$ for some $p \in C$ and so $T_2^{n_k+1} x_{n_k} \to T_2 p$ as $k \to \infty$. From (34), we have $x_{n_k} \to p$ as $k \to \infty$. Also, by (35) we get that $T_1^{n_k} y_{n_k} \to p$ as $k \to \infty$. Consider

$$d(x_{n_k+1}, x_{n_k}) = d\left((1-\alpha_{n_k})x_{n_k} \oplus \alpha_{n_k} T_1^{n_k} y_{n_k}, x_{n_k}\right) \le d\left(x_{x_k}, T_1^{n_k} y_{n_k}\right). \tag{36}$$

From (35) and (36), it follows that $x_{n_k+1} \to p$ as $k \to \infty$. Again, from (35), we have $T_1^{n_k+1} y_{n_k} \to T_1 p$.

Next, we show that $p \in \mathbb{F}$. Notice that

$$d(p, T_2 p) \le d(p, x_{n_k+1}) + d\left(x_{n_k+1}, T_2^{n_k+1} x_{n_k+1}\right) + d\left(T_2^{n_k+1} x_{n_k+1}, T_2^{n_k+1} x_{n_k}\right) + d\left(T_2^{n_k+1} x_{n_k}, T_2 p\right) \tag{37}$$

Since T_2 is uniformly continuous, taking the limit as $k \to \infty$, and using (34) we obtain that $p = T_2 p$ and so $p \in F(T_2)$. Notice that

$$d(p, T_1p) \le d(p, x_{n_k+1}) + d\left(x_{n_k+1}, T_1^{n_k+1}x_{n_k+1}\right) \\ +d\left(T_1^{n_k+1}x_{n_k+1}, T_1^{n_k+1}x_{n_k}\right) + d\left(T_1^{n_k+1}x_{n_k}, T_1p\right). \tag{38}$$

Letting $k \to \infty$, we also obtain that $p = T_1p$ and hence $p \in F(T_1)$. Therefore $p \in \mathbb{F}$. Hence, by Lemma 2.6, $x_n \to p \in \mathbb{F}$, since $\lim_{n\to\infty} d(x_n, p)$ exists. If T_1 is compact, then essentially the same arguments as above our result follow. This completes the proof.

We give the following example in support of our main results.

Example 3.5. Let $\mathbb{X} = \mathbb{R}^1$. and $C = [0, 1]$, a closed convex subset of $\mathbb{X}$ and define $T_1, T_2 : C \to C$ by

$$T_1x = \begin{cases} \dfrac{x}{2} & , \text{if } x \in \left[0, \dfrac{1}{2}\right] \\ 0 & , \text{if } x \in \left(\dfrac{1}{2}, 1\right] \end{cases}$$

and

$$T_2x = \begin{cases} x & , \text{if } x \in \left[0, \dfrac{1}{2}\right] \\ \dfrac{1}{2} & , \text{if } x \in \left(\dfrac{1}{2}, 1\right]. \end{cases}$$

Then, T_1, T_2 are asymptotically quasi-nonexpansive but not nonexpansive with $\mathbb{F} = \{0\} \neq \emptyset$. For

$$d(T_1^n x, T_1^n y) \le \frac{1}{2^n} d(x, y) \le d(x, y), \forall x, y \in \left[0, \frac{1}{2}\right].$$

And

$$d(T_1^n x, T_1^n y) = 0 \le d(x, y), \forall x, y \in \left(\frac{1}{2}, 1\right].$$

Hence, T_1 is asymptotically quasi-nonexpansive. Similarly, we can show that T_2 is asymptotically quasi-nonexpansive.

Define a sequence $\{x_n\}$ as in (6) by starting from arbitrary $x_1 \in C$,

$$\begin{aligned} x_{n+1} &= (1 - \alpha_n)x_n + \alpha_n T_1^n y_n \\ y_n &= (1 - \beta_n)x_n + \beta_n T_2^n x_n, \end{aligned} \tag{39}$$

for all $n \ge 1$, where $\{\alpha_n\}$ and $\{\beta_n\}$ are sequences in $[0, 1]$. Taking $\alpha_n = \frac{1}{2} = \beta_n$. Next, we construct a sequence $\{x_n\}$. Starting from $x_1 = 1$, we get

$$y_1 = \frac{1}{2}x_1 + \frac{1}{2}T_2x_1 = \frac{1}{2}(1) + \frac{1}{2}T_2(1) = \frac{1}{2}(1) + \frac{1}{2}\left(\frac{1}{2}\right) = \frac{3}{4},$$

we get

$$x_2 = \frac{1}{2}x_1 + \frac{1}{2}T_1y_1 = \frac{1}{2}(1) + \frac{1}{2}T_1\left(\frac{3}{4}\right) = \frac{1}{2} + \frac{1}{2}(0) = \frac{1}{2} = 0.5.$$

$$y_2 = \frac{1}{2}x_2 + \frac{1}{2}T_2^2x_2 = \frac{1}{2}\left(\frac{1}{2}\right) + \frac{1}{2}T_2^2\left(\frac{1}{2}\right) = \frac{1}{4} + \frac{1}{2}\left(\frac{1}{2}\right) = \frac{1}{2},$$

we get

$$x_3 = \frac{1}{2}x_2 + \frac{1}{2}T_1^2y_2 = \frac{1}{2}\left(\frac{1}{2}\right) + \frac{1}{2}T_1^2\left(\frac{1}{2}\right) = \frac{1}{4} + \frac{1}{2}\left(\frac{1}{8}\right) = \frac{5}{16} = 0.3125.$$

$$y_3 = \frac{1}{2}x_3 + \frac{1}{2}T_2^3x_3 = \frac{1}{2}\left(\frac{5}{16}\right) + \frac{1}{2}T_2^3\left(\frac{5}{16}\right) = \frac{5}{32} + \frac{1}{2}\left(\frac{5}{16}\right) = \frac{5}{16},$$

we get

$$x_4 = \frac{1}{2}x_3 + \frac{1}{2}T_1^3y_3 = \frac{1}{2}\left(\frac{5}{16}\right) + \frac{1}{2}T_1^3\left(\frac{5}{16}\right) = \frac{5}{32} + \frac{1}{2}\left(\frac{5}{128}\right) = \frac{45}{256} = 0.1757.$$

$$y_4 = \frac{1}{2}x_4 + \frac{1}{2}T_2^4x_4 = \frac{1}{2}\left(\frac{45}{256}\right) + \frac{1}{2}T_2^4\left(\frac{45}{256}\right) = \frac{45}{512} + \frac{1}{2}\left(\frac{45}{256}\right) = \frac{45}{256},$$

we get

$$x_5 = \frac{1}{2}x_4 + \frac{1}{2}T_1^4y_4 = \frac{1}{2}\left(\frac{45}{256}\right) + \frac{1}{2}T_1^4\left(\frac{45}{256}\right) = \frac{45}{512} + \frac{1}{2}\left(\frac{45}{4096}\right) = \frac{765}{8192} = 0.0933.$$

Proceeding in a similar method, we will get a sequence $\{x_n\}$ that converges to 0, the common fixed point of T_1 and T_2, that is, we obtain the sequence

$$1, \frac{1}{2}, \frac{5}{16}, \frac{45}{256}, \frac{765}{8192}, \ldots, x_n \to 0.$$

Corollary 3.6. *Let $\mathbb{X}$ be a CAT(0) space and C a nonempty* **compact** *convex subset of* $\mathbb{X}$. *Let $T_1, T_2 : C \to C$ be two uniformly continuous asymptotically quasi-nonexpansive mappings, and $\mathbb{F} = F(T_1) \cap F(T_2) \neq \varnothing$. Let $\{\alpha_n\}$ and $\{\beta_n\}$ be sequences in $[\varepsilon, 1-\varepsilon]$ forsome $\varepsilon \in (0, 1)$. From arbitrary $x_1 \in C$, define the sequence $\{x_n\}$ by iteration (6). Assume, in addition, that either T_2 or T_1 is compact. Then, $\{x_n\}$ converges strongly to some common fixed point of T_1 and T_2.*

Corollary 3.7. *Let $\mathbb{X}$ be a CAT(0) space and C a nonempty* **compact** *convex subset of* $\mathbb{X}$. *Let $T : C \to C$ be two uniformly continuous asymptotically quasi-nonexpansive mappings with sequences $\{k_n\} \subset [0, \infty)$ such that $\sum_{n=1}^{\infty} k_n < \infty$. Let $\{\alpha_n\}$ and $\{\beta_n\}$ be sequences in $[\varepsilon, 1-\varepsilon]$ forsome $\varepsilon \in (0, 1)$. From arbitrary $x_1 \in K$, define the sequence $\{x_n\}$ by the iteration*

$$\begin{aligned} x_{n+1} &= (1-\alpha_n)x_n \oplus \alpha_n T^n y_n \\ y_n &= (1-\beta_n)x_n \oplus \beta_n T^n x_n, \end{aligned} \tag{40}$$

with $n \geq 1$. Then, $\{x_n\}$ converges strongly to some fixed point of T.

Corollary 3.8. *Let* X *be a* **Hibert** *space and* C *a nonempty closed convex subset of* X. *Let* $T_1, T_2 : C \to C$ *be two uniformly continuous asymptotically quasi-nonexpansive mappings, and* $\mathbb{F} = F(T_1) \cap F(T_2) := \{x \in K : T_1 x = T_2 x = x\} \neq \emptyset$. *Let* $\{\alpha_n\}$ *and* $\{\beta_n\}$ *be sequences in* $[\varepsilon, 1 - \varepsilon]$ *forsome* $\varepsilon \in (0, 1)$. *From arbitrary* $x_1 \in C$, *define the sequence* $\{x_n\}$ *by the iterative scheme* (6). *Assume, in addition, that either* T_2 *or* T_1 *is compact. Then,* $\{x_n\}$ *converges strongly to some common fixed point of* T_1 *and* T_2.

4. Conclusions

In this chapter, we establish strong convergence results for two asymptotically quasi-nonexpansive mappings T_1, T_2 in the setting of CAT(0) spaces via the sequence $\{x_n\}$ generated iteratively by arbitrary $x_1 \in C$, $x_{n+1} = (1 - \alpha_n)x_n \oplus \alpha_n T_1^n y_n$, $y_n = (1 - \beta_n)x_n \oplus \beta_n T_2^n x_n, n \geq 1$. We obtained the following results:-.

a. Lemma 3.1, an extension of Theorem 1.2 (See [24], Theorem 3.1).

b. Theorem 3.2, it is proved that the sequence $\{x_n\}$ converges strongly to a common fixed point of T_1 and T_2 if and only if $\lim_{n \to \infty} d(x_n, \mathbb{F}) = 0$. (Note that T_1 and T_2 need not be continuous). This theorem extends and improves Theorem 1.3 (See [24] Theorem 3.2).

c. Lemma 3.3, it is proved that

$$\lim_{n \to \infty} d(x_n, T_2^n x_n) = \lim_{n \to \infty} d(x_n, T_1^n x_n) = \lim_{n \to \infty} d(x_n, T_1^n y_n) = 0.$$

   This lemma extends and improves Theorem 3.3 in [24].

d. Theorem 3.4, it is proved that If $T_1, T_2 : C \to C$ be two uniformly continuous asymptotically quasi-nonexpansive mappings. Suppose, in addition, that either T_2 or T_1 is compact. Then, $\{x_n\}$ converges strongly to some common fixed point of T_1 and T_2. This theorem significantly extends and improves Theorem 1.3 (See [24], Theorem 3.4).

As consequence, we obtain Corollaries 3.6, 3.7, and 3.8. All of our results remain true for the subclass of asymptotically nonexpansive mappings.

Acknowledgements

The authors would like to thank the referee for his/her useful comments on the improvement of this chapter. The second author wishes to thank the faculty of Science of Maejo University for moral support in the writing of this manuscript.

Competing interests

The authors declare that they have no competing interests.

Author details

Jamnian Nantadilok[1*†] and Buraskorn Nuntadilok[2†]

1 Faculty of Science, Department of Mathematics, Lampang Rajabhat University, Lampang, Thailand

2 Faculty of Science, Department of Mathematics, Maejo University, Chiang Mai, Thailand

*Address all correspondence to: jamnian2010@gmail.com

† These authors contributed equally.

References

[1] Dotson WG Jr. Fixed points of quasi-nonexpansive mappings. Australian Mathematical Society A. 1972;**13**:167-170

[2] Goebel K, Kirk WA. A fixed point theorem for asymptotically nonexpansive mappings. Proceedings of the American Mathematical Society. 1972;**35**:171-174

[3] Goebel K, Kirk WA. A fixed point theorem for transformations whose iterates have uniform Lipschitz constant, Polska Akademia Nauk. Instytut Matematyczny. Studia Mathematica. 1973;**47**:135-140

[4] Chidume CE. Iterative algorithms for nonexpansive mappings and some of their generalizations. In: Agarwal RP, et al, editors. Nonlinear Analysis and Applications: to V. Lakshmikantham on His 80th Birthday. Vol. 1, 2. Dordrecht: Kluwer Academic; 2003, pp. 383-429

[5] Chidume CE, Ofoedu EU, Zegeye H. Strong and weak convergence theorems for asymptotically nonexpansive mappings. Journal of Mathematical Analysis and Applications. 2003;**280**(2): 364-374

[6] Fukhar-ud-din H, Khan SH. Convergence of iterates with errors of asymptotically quasi-nonexpansive mappings and applications. Journal of Mathematical Analysis and Applications. 2007;**328**:821-829

[7] Ghosh MK, Debnath L. Convergence of Ishikawa iterates of quasi-nonexpansive mappings. Journal of Mathematical Analysis and Applications. 1997;**207**(1):96-103

[8] Ishikawa S. Fixed points by a new iteration method. Proceedings of the American Mathematical Society. 1974; **44**:147-150

[9] Khan SH, Hussain N. Convergence theorems for nonself asymptotically nonexpansive mappings. Computers & Mathematics with Applications. 2008; **55**(11):2544-2553

[10] Mann WR. Mean value methods in iteration. Proceedings of the American Mathematical Society. 1953;**4**:506-510

[11] Petryshyn WV, Williamson TE Jr. Strong and weak convergence of the sequence of successive approximations for quasi-nonexpansive mappings. Journal of Mathematical Analysis and Applications. 1973;**43**:459-497

[12] Qihou L. Iterative sequences for asymptotically quasi-nonexpansive mappings. Journal of Mathematical Analysis and Applications. 2001;**259**(1):1-7

[13] Qihou L. Iterative sequences for asymptotically quasi-nonexpansive mappings with error member. Journal of Mathematical Analysis and Applications. 2001;**259**(1):18-24

[14] Rhoades BE. Fixed point iterations for certain nonlinear mappings. Journal of Mathematical Analysis and Applications. 1994;**183**(1):118-120

[15] Schu J. Weak and strong convergence to fixed points of asymptotically nonexpansive mappings. Bulletin of the Australian Mathematical Society. 1991;**43**(1):153-159

[16] Senter HF, Dotson WG. Approximating fixed points of nonexpansive mappings. Proceedings of the American Mathematical Society. 1974;**44**:375-380

[17] Tan KK, Xu HK. Approximating fixed points of nonexpansive mappings

by the Ishikawa iteration process. Journal of Mathematical Analysis and Applications. 1993;**178**(2):301-308

[18] Wang L. Strong and weak convergence theorems for common fixed point of nonself asymptotically nonexpansive mappings. Journal of Mathematical Analysis and Applications. 2006;**323**(1):550-557

[19] Yang L. Modified multistep iterative process for some common fixed point of a finite family of nonself asymptotically nonexpansive mappings. Mathematical and Computer Modelling. 2007; **45**(9–10):1157-1169

[20] Zhou H, Agarwal RP, Cho YJ, Kim YS. Nonexpansive mappings and iterative methods in uniformly convex Banach spaces. Georgian Mathematical Journal. 2002;**9**(3):591-600

[21] Zhou HY, Cho YJ, Kang SM. A new iterative algorithm for approximating common fixed points for asymptotically nonexpansive mappings. Fixed Point Theory and Applications. 2007;**2007**:10

[22] Khan SH, Takahashi W. Approximating common fixed points of two asymptotically nonexpansive mappings. Scientiae Mathematicae Japonicae. 2001;**53**(1):143-148

[23] Qihou L. Iteration sequences for asymptotically quasi-nonexpansive mapping with an error member of uniform convex Banach space. Journal of Mathematical Analysis and Applications. 2002;**266**(2):468-471

[24] Shahzad N, Udomene A. Approximating common fixed points of two asymptotically quasi-nonexpansive mappings in Banach spaces. Fixed Point Theory and Applications. 2006:1-10. DOI: 10.1155/FPTA/2006/18909

[25] Bridson M, Haefliger A. Metric Spaces of Non-Positive Curvature. Berlin: Springer; 1999

[26] Brown KS. Buildings. New York: Springer; 1989

[27] Goebel K, Reich S. Uniform Convexity, Hyperbolic Geometry, and Nonexpansive Mappings. New York: Marcel Dekker, Inc.; 1984

[28] Bruhat F, Tits J. Groupes réductifs sur un corps local. I. Données radicielles valuées. Institut des Hautes études Scientifiques. 1972;**41**:5-251 (in French)

[29] Amnuaykarn K, Kumam P, Nantadilok J. On the existece of best proximity points of multi-valued mappings in CAT(0) spaces. Journal of Nonlinear Functional Analysis. 2021:1-14

[30] Dhompongsa S, Kirk WA, Panyanak B. Nonexpansive set-valued mappings in metric and Banach spaces. Journal of Nonlinear and Convex Analysis. 2007;**8**: 35-45

[31] Dhompongsa S, Panyanak B. On Δ-convergence theorems in CAT(0) spaces. Computers & Mathematcs with Applications. 2008;**56**:2572-2579

[32] Nantadilok J, Khanpanuk C. Best proximity point results for cyclic contractions in CAT(0) spaces. Computers & Mathematcs with Applications. 2021;**12**(2):1-9

[33] Nanjarus B, Panyanak B. Demicloesed principle for asymptotically nonexpansive mappings in CAT(0) spaces. Fixed Point Theory and Applications. 2010:11. DOI: 10.1155/2010/268780

[34] Sastry KPR, Babu GVR. Convergence of Ishikawa iterates for a multi-valued mapping with a fixed

point. Czechoslovak Mathematical Journal. 2005;**55**:817-826

[35] Kirk WA. Fixed point theory for nonexpansive mappings II. Contemporary Mathematics. 1983;**18**: 121-140

[36] Leustean L. A quadratic rate of asymptotic regularity for CAT(0) spaces. Journal of Mathematical Analysis and Applications. 2007;**325**(1):386-399

[37] IbnDehaish BA, Khamsi MA, Khan AR. Mann iteration process for asymptotic pointwise nonexpansive mappings in metric spaces. Journal of Mathematical Analysis and Applications. 2013;**397**:861-868

[38] Fukhar-ud-din H, Khan AR, Akhtar Z. Fixed point results for ageneralized nonexpansive map in uniformly convex metric spaces. Nonlinear Analysis. 2012; **75**:4747-4760

[39] Fukhar-ud-din H, Khamsi MA. Approximating common fixed points in hyperbolic spaces. Fixed Point Theory and Applications. 2014;**2014**:113

Chapter 4

Iterative Algorithms for Common Solutions of Nonlinear Problems in Banach Spaces

Getahun Bekele Wega

Abstract

The purpose of this manuscript is to construct an iterative algorithm for approximating a common solution of variational inequality problem and g-fixed point problem of pseudomonotone and Bregman relatively g-nonexpansive mappings, respectively, and prove strong convergence of a sequence generated by the proposed method to a common solution of the problems in real reflexive Banach spaces. The assumption that the mapping is Lipschitz monotone mapping is dispensed with. In addition, we give an application of our main result to find a minimum point of a convex function in real reflexive Banach spaces. Finally, we provide a numerical example to validate our result. Our results extend and generalize many results in the literature.

Keywords: common solution, Bregman relatively g-nonexpansive, g-fixed point, monotone mapping, pseudomonotone mapping, variational inequality

1. Introduction

Let E be a real Banach space with its dual space E^*. Let C be a nonempty, closed, and convex subset of E. A mapping $G : C \to E^*$ is said to be monotone provided that for all points p and q in C,

$$\langle Gp - Gq, p - q \rangle \geq 0. \tag{1}$$

It is called α-strongly monotone if there exists a positive real number α such that for all points p and q in C,

$$\langle Gp - Gq, p - z \rangle \geq \ \|Gp - Gq\|^2. \tag{2}$$

We remark that α-strongly monotone is α^{-1}–Lipschitz monotone mapping. A mapping $G : C \to E^*$ is called pseudomonotone mapping provided that for all points p and q in C,

$$\langle Gp, p - q \rangle \geq 0 \ \text{ implies } \ \langle Gq, p - q \rangle \geq 0. \tag{3}$$

From inequalities (1)-(3) above, we can observe that the class of pseudomonotone mappings contains the classes of monotone and α-strongly monotone mappings. Let $G : C \to E^*$ be a mapping. The variational inequality problem (VIP) introduced by Hartman and Stampacchia [1] in 1966 is mathematically formulated as the problem of finding a point z in C such that for all points p in C,

$$\langle Gz, p - z \rangle \geq 0. \tag{4}$$

We denote the solution set of problem (4) by $VIP(C, V)$. This problem contains, as special cases, many problems in the fields of applied mathematics, such as mechanics, physics, engineering, the theory of convex programming, and the theory of control. Consequently, considerable research efforts have been devoted to methods of finding approximate solutions of variational inequality problems in several directions for different classes of mappings (see, e.g., [2–10]).

Several authors have also studied, different iterative algorithms for approximating a common solution of VIP and fixed point problem of Lipschitz monotone and nonexpansive mappings, respectively (see, e.g., [4, 7, 11–15]).

In 2003, Takahashi and Tododa [13] introduced an iterative algorithm for finding a common solution for VIP and fixed point problem of α-strongly monotone and nonexpansive mappings, respectively, in Hilbert spaces setting. Under certain conditions, they proved that the sequence generated by their proposed method converges weakly to a common solution.

In 2005, Iiduka and Takashi [3] studied an iterative scheme for finding a common solution of VIP and fixed point problem of α-strongly monotone and nonexpansive mappings, respectively, in Hilbert spaces setting. They proved that the sequence generated by their proposed scheme converges strongly to a common solution provided that the control sequences satisfy appropriate conditions.

In 2016, Zhang and Yuan [16] established an algorithm for approximating a common solution of VIP and fixed point problem for a finite family of α-inverse strongly monotone and nonexpansive mappings, respectively, in the Hilbert spaces setting. They proved strong convergence of the sequence proposed by their method.

In space, more general than Hilbert spaces, Tufa and Zegeye [17] introduced an iterative algorithm for approximating a common solution of VIP and fixed point problem of Lipschitz monotone and relatively nonexpansive mappings, respectively in real 2-uniformly convex and uniformly smooth Banach spaces. They proved that the sequence generated by their algorithm converges strongly to a common solution of the problems. A mapping $T : C \to E^*$ is said to be relatively nonexpansive if $F(T) \neq \emptyset$, $\phi(z, Tu) \leq \phi(z, u)$ $\forall\, u \in C$, $z \in F(T)$ and $\hat{F}(T) = F(T)$, where $F(T)$, is the set of fixed points of T and $\hat{F}(T)$ is the set of asymptotical fixed point of T.

Recently, Wega and Zegeye [18] introduced an iterative scheme for approximating a common solution of VIP and g-fixed point problem (GFP) of Lipschitz monotone and Bregman relatively g-nonexpansive mappings, respectively in real reflexive Banach spaces and obtained strong convergence results. A mapping $T : C \to E^*$ is said to be Bregman relatively g-nonexpansive (BRGN) if $F(T) \neq \emptyset$, $D_g(z, Tu) \leq D_g(z, u)$ $\forall\, u \in C$, $z \in F(T)$ and $\hat{F}_g(T) = F_g(T)$, where $F_g(T)$, is the set of g-fixed points of T and $\hat{F}_g(T)$ is the set of asymptotical g-fixed point of T, where g is a convex function of E satisfies certain conditions. A point z in C is said to be g-fixed point of T provided that $Tz = \nabla\, gz$.

Motivated and inspired by the above results, it is our purpose in this book chapter to construct an iterative algorithm, which converge strongly to a common element of the set of VIP solutions of continuous pseudomonotone and the set GFPP of BRGN mappings in real reflexive Banach spaces. In addition, we give an application of our main result to find a minimum point of a convex function and provide a numerical example to validate our main result. Our results extend and generalize many results in the literature.

Now, we recall some definitions that we will need in the sequel.

Hereafter in this paper let E be a real reflexive Banach space with its dual space E^*, C be a nonempty, convex and closed subset of E and let $\mathcal{G}$ be a family of proper, lower semi-continuous and convex functions on E.

Let g be an element of $\mathcal{G}$. The domain of g, $dom\ \ g$, is given by $dom\ \ g = \{p \in E : g(p) < \infty\}$, the Fenchel conjugate of g at p^*, $g^*(p^*)$, is given by $g^*(p^*) = sup\{\langle p^*,\ p\rangle - g(p) : p \in E\ \ and\ \ p^* \in E^*\}$, the subdifferential of g at p, $\partial g(p)$, is given by $\partial g(p) = \{p^* \in E^* : g(q) \geq g(p) + \langle p^*,\ q - p\rangle, \forall p \in E\}$, the right-hand derivative of g at u in the direction of q, $g'(p, q)$, is given by:

$$g'(p, q) = \lim_{s \to 0_+} \frac{g(p + sq) - g(p)}{s}, \tag{5}$$

and the gradient of g, at p is a linear function, $\nabla\ g$, is given by $\langle \nabla g(p), q\rangle = g'(p, q)$.
The function g is called:

i. Gâteaux differentiable at p element of E if the limit in (5) exists for any q in E as $s \to 0$.

ii. Gâteaux differentiable if it is Gâteaux differentiable at every element u in $int\ \ domg$.

iii. Uniformity Fréchet differentiable on C if the limit as $s \to 0$ in (5) attained uniformly for $p \in C$ and $\|q\| = 1$.

iv. Strongly coercive if $\lim_{\|p\| \to \infty} \frac{g(p)}{\|p\|} = \infty$.

Gâteaux differentiable function g is called Legendre if g^* is Gâteaux differentiable, both $int\ \ dom\ \ g$ and $int\ \ dom\ \ g^*$ are nonempty, $dom\ \ \nabla\ g = int\ \ dom\ \ g$ and $dom\ \ \nabla\ g^* = int\ \ dom\ \ g^*$.

Remark 1.1 $\nabla\ g^* = (\nabla\ g)^{-1}$ (see, [19]) provided that g is Legendre function and the gradient of Legendre function g defined by $g(u) = \frac{\|u\|^p}{p}$ is coinciding with the generalized duality map, that is, $\nabla g = J_p$, where $(1 < p, q < \infty)$ and q is a conjugate of p (see, e.g., [20]).

The Bregman distance with respect to g (see, e.g., [21]) is a function D_g : $dom\ \ g \times int\ \ dom\ \ g \to [0, \infty)$ defined by:

$$D_g(q, p) = g(q) - g(p) - \langle \nabla g(p), q - p\rangle, \tag{6}$$

where g is Gâteaux differentiable. The Bregman projection with respect to g at p in $int\ \ dom\ \ g$ onto C is denoted by $P^g_C p$ defined by $D_g(P^g_C p, p) = \inf\{D_g(q,\ p) : \forall q \in C\}$.

Remark 1.2 We note that the Bregiman distance is not distance in the usual sense. However, it has the following properties (see, e.g., [22–24]):

i. The three point identity:

$$D_g(p,q) + D_g(q,w) - D_g(p,q) = \langle \nabla\ g(w) - \nabla\ g(q), p - q \rangle \tag{7}$$

for all $q \in dom\ \ g$ and $p, w \in int\ \ dom\ \ g$.

ii. The four point identity:

$$D_g(q,p) + D_g(q,z) - D_g(w,p) + D_g(w,z) - \langle \nabla\ g(z) - \nabla\ g(p), q - w \rangle, \tag{8}$$

for all $q, w \in dom\ \ g$ and $p, z \in int\ \ dom\ \ g$.

Lemma 1.3 Let g be a totally convex and Gáteaux differentiable on *int domg*. Let $p \in int\ \ domg$. Then, the P_c^g from E onto C is a unique point with the following properties [25]:

i. $\langle \nabla\ g(p) - \nabla\ g(z), q - z \leq 0$ if and only if $z = P_C^g p$, $\forall q \in C$.

ii. $D_g(p,q) \geq D_g(q, P_C^g p) + D_g(P_C^g p, p)$, $\forall q \in C$.

Let g be a Legendre and $V_g : E \times E^* \to [0, \infty)$ be a function defined by:

$$V_g(p,p^*) = g(p) - \langle p^*, p \rangle + \nabla\ g^*(q^*), \forall p \in E, p^* \in E^*. \tag{9}$$

Then, V_g is nonnegative which satisfies (see, e.g., [26])

$$V_g(p,p^*) = D_g(p, \nabla\ g^*(p^*)) \tag{10}$$

and

$$V_g(p,p^*) \leq V_g(p, p^* + q^*) - \langle q^*, \nabla\ g^*(p^*) - p \rangle, \tag{11}$$

for all $p \in E$ and $p^* \in E^*$.

Lemma 1.4 If g is lower, convex, semi-convex proper function, then g^* is a weak* lower semi-convex and proper function and hence, we have

$$D_g\left(w, \nabla\ g^*\left(\sum_{i=1}^{N} s_i \nabla g(p_i)\right)\right) \leq \sum_{i=1}^{N} s_i D_g(w, p_i), \tag{12}$$

for all w in E, where $\{p_i\} \subseteq E$ and $\{s_i\} \subseteq (0,1)$ with $\sum_{i=1}^{N} s_i = 1$ [27].

A Gâteaux differentiable function g is called.

i. Uniformly convex function (see, [28]), provided that for all p and q *dom g* $s \in [0,1]$, we have

$$g(sp + (1-s)q) \leq sg(p) + (1-s)g(p) - (1-s)s\phi(\|p - q\|), \tag{13}$$

where ϕ is a function that is increasing and vanishes only at zero.

i. Strongly convex with constant $\alpha > 0$ for all u and q elements of $domg$ (see, [29])

$$\langle \nabla\ g(p) - \nabla\ g(p), p - q \rangle \geq \alpha \|p - q\|^2. \tag{14}$$

ii. Totaly convex if $\nu_g(p, s) = \inf_{\{p \in E: \|p-q\|=s\}} D_g(q, p) > 0$, for all $p \in E$ and $s > 0$.

We note that g is uniformly convex if and only if g is totally convex on bounded subsets of E (see, [25], Theorem 2.10 p. 9). Moreover, the class of uniformly convex function functions contains the class of strongly convex functions.

Lemma 1.5 Let E be a Banach space and $r > 0$ be a constant. Let $g : E \to \mathbb{R}$ be a continuous convex function that is uniformly convex on bounded subsets of E. Then,

$$g\left(\sum_{k=0}^{n} \beta_k u_k\right) \leq \sum_{k=0}^{n} \beta_k g(u_k) - \beta_i \beta_j \rho_r(\|u_i - u_j\|), \tag{15}$$

$\forall 0 \leq i,j \leq n$, $u_k \in B_r$, $\beta_k \in (0, 1)$ with $\sum_{k=0}^{n} \beta_k = 1$, where ρ_r is the gauge of uniform convexity of g [30].

Lemma 1.6 Let g be a total convex G$\hat{a}$ teaux differentiable such that $dom\ \ g = E$. Then, for each $x^* \in E^*\ \ \{0\}, \tilde{y} \in E, x \in H^+$ and $\tilde{x} \in H^-$, it holds that

$$D_g(\tilde{x}, x) \geq D_g(\tilde{x}, z) + D_g(z, x), \tag{16}$$

where $z = argmin_{y \in H} D_g(y, x)$ and $H^- = \{y \in E : \langle x^*,\ y - \tilde{y} \rangle \leq 0\}$, $H = \{y \in E : \langle x^*,\ y - \tilde{y} \rangle = 0\}$ and $H^+ = \{y \in E : \langle x^*,\ y - \tilde{y} \rangle \geq 0\}$.

2. An iterative algorithm for a common solution of variational inequality and $g-$fixed problems

In this section, let E be a real reflexive Banach space with its dual space E^*. Let C be a nonempty, closed, and convex subset of E. Let $g : E \to (-\infty, +\infty] \in \mathcal{G}$ be a uniformly Fr$\hat{e}$chet differentiable Legendre which is bounded, uniformly convex, and strongly coercive on bounded subsets of E. We denote the family of such functions by $\mathcal{G}(E)$.

In the sequel, we shall make use of the following assumptions.

Assumption:

A1) Let $l \in (0, 1)$, $\mu > 0$ and $\left[\beta \in \underline{\beta}, \overline{\beta}\right] \subset \left(0, \frac{1}{\mu}\right)$.

A2) Let $\{\alpha_n\} \subset (0, c)$ with the properties $\lim_{n\to\infty} \alpha_n = 0\ \ and\ \ \sum_{n=1}^{\infty} \alpha_n = \infty$, where $c > 0$.

Algorithm 1: For any $x_0, v \in C$, define an algorithm by.

Step 1. Compute

$$y_n = \nabla\ g^*\left[\nabla\ gx_n - \beta G x_n\right] \text{ and } d(y_n) = x_n - P_C^g y_n. \tag{17}$$

If $d(y_n) = 0$ and $\nabla\ gx_n - Tx_n = 0$, then stop and $x_n \in \Omega$. Otherwise,

---

Step 2. Compute $p_n = x_n - \tau_n d(y_n)$,

where $\tau_n = l^{j_n}$ and j_n is the smallest nonnegative integer j satisfying

$$\langle Gx_n - Gp_n,\ d(y_n)\rangle \leq \mu D_g(P_C^g y_n,\ x_n). \tag{18}$$

Step 3. Compute

$$\begin{cases} a_n = P_{P_n}^f \nabla g^*(\nabla gx_n - \beta Gp_n), \\ r_n = \nabla g^*(\eta_{n,1}\nabla gx_n + \eta_{n,2}Tx_n + \eta_{n,3}\nabla gu_n), \\ x_{n+1} = P_C^g \nabla g^*(\alpha_n \nabla gv + (1-\alpha_n)\nabla gr_n), \end{cases} \tag{19}$$

where $g \in \mathcal{G}(E)$, $P_n = \{p \in C : \langle Gp_n,\ p - p_n\rangle = 0\}$, $u_n = P_C^g a_n$ and $\{\eta_{n,i}\} \subset [\varepsilon, 1) \subset (0, 1)$, for $i = 1,2,3$ such that $\sum_{i=1}^{3}\eta_{n,i} = 1$, $\forall n \geq 0$.

Step 4. Set $n := n+1$ and go to **Step 1**.

We shall need the following Lemmas in the sequel.

Lemma 1.7 Assume that $\{x_n\}$ and $\{y_n\}$ are sequences generated by Algorithm 1. Then, the search rule in Step 2 is well defined.

Proof: Since $l \in (0, 1)$ and G is continuous on C, we have

$$\langle Gx_n - Gp_n, d(y_n)\rangle \to 0 \tag{20}$$

as $j \to \infty$. On the other hand, the fact that $D_g(P_C^g y_n, x_n) > 0$, there exists a nonnegative integer j_n satisfying the inequality in Step 2, and the claim holds.

Lemma 1.8 Assume that $\{x_n\}$ and $\{y_n\}$ are sequences generated by Algorithm 1. Then, we have:

$$\langle Gx_n, d(y_n)\rangle \geq \frac{1}{\beta} D_g(P_C^g y_n, x_n) \tag{21}$$

Proof: From (17), we have:

$$\nabla gy_n = \nabla gx_n - \beta Gx_n, \tag{22}$$

which implies:

$$\nabla gx_n - \nabla gy_n = \beta Gx_n. \tag{23}$$

Thus, from (23), (17), and (7), we get:

$$\langle Gx_n, d(y_n)\rangle = \frac{1}{\beta}\langle \nabla gx_n - \nabla gy_n, x_n - P_C^g y_n\rangle \tag{24}$$

$$= \frac{1}{\beta}\left[D_g(P_C^g y_n,\ x_n) + D_g(x_n,\ y_n) - D_g(P_C^g y_n,\ y_n)\right] \tag{25}$$

$$\geq \frac{1}{\beta} D_g(P_C^g y_n, x_n), \tag{26}$$

and hence the assertion hold.

Lemma 1.9 Suppose the assumption (A1) holds. Let $G : C \to E^*$ be a continuous pseudomonotone mapping. Then, $\langle Gp_n, x_n - p_n \rangle \geq \tau_n \left(\frac{1}{\beta} - \mu\right) D_g\left(P_C^g y_n, x_n\right)$. In particular, if $d(y_n) \neq 0$, then $\langle Gp_n, x_n - p_n \rangle > 0$.

Proof: Using Step 2 of the algorithm we know that

$$\langle Gp_n, x_n - p_n \rangle = \langle Gp_n, x_n - (x_n - \tau_n d(y_n)) \rangle \tag{27}$$

$$= \tau_n \langle Gp_n, d(y_n) \rangle. \tag{28}$$

On the other hand, from (18), we have:

$$\langle Gx_n - Gp_n, d(y_n) \rangle \leq \mu D_g\left(P_C^g y_n, x_n\right) \tag{29}$$

which implies that

$$\langle Gp_n, d(y_n) \rangle \geq \langle Gx_n, d(y_n) \rangle - \mu D_g\left(P_C^g y_n, x_n\right). \tag{30}$$

From (30) and Lemma 8, we get:

$$\langle Gp_n, d(y_n) \rangle \geq \left(\frac{1}{\beta} - \mu\right) D_g\left(P_C^g y_n, x_n\right). \tag{31}$$

Combining (28) and (31), we obtain:

$$\langle Gp_n, x_n - p_n \rangle \geq \tau_n \left(\frac{1}{\beta} - \mu\right) D_g\left(P_C^g y_n, x_n\right), \tag{32}$$

and the proof is complete.

Theorem 1.10 Suppose the Assumptions (A1) and (A2) hold. Let $G : C \to E^*$ and $T : C \to E^*$ be continuous pseudomonotone and BRGN mappings, respectively, with $\Omega = VI(C, G) \cap F_g(T) \neq \emptyset$. Then, the sequens $\{x_n\}$ generated by Algorithm 1 is bounded.

Proof: Let $x^* = P_\Omega^g(v)$ and $w_n = \nabla g^*\left(\alpha_n \nabla gv + (1 - \alpha_n) \nabla gr_n\right)$. We note that from Lemma 1.3 (i), we obtain

$$\langle u - x^*, \nabla gv - \nabla gx^* \rangle \leq 0, \forall u \in \Omega. \tag{33}$$

Now, for each $n \geq 0$, define the sets: $P_n^- = \{p \in C : \langle Gx_n, p - x_n \rangle \leq 0\}$, $P_n = \{p \in C : \langle Gx_n, p - x_n \rangle = 0\}$, and $P_n^+ = \{p \in C : \langle Gx_n, p - x_n \rangle \geq 0\}$. Let $x^* \in \Omega$, from definition of G, we have $\langle Gx^*, y - x^* \rangle \geq 0$, which implies that $\langle Gy, y - x^* \rangle \geq 0$ for all $y \in C$, and hence, $x^* \in P_n^-$ for all $n \geq 0$. Moreover, from Lemma 9, we have $\langle Gp_n, x_n - p_n \rangle > 0$, which implies that $x_n \in P_n^+$ and $x_n \notin P_n^-$ for all $n \geq 0$. Now, from Lemma 1.6, we get:

$$D_g(x^*, a_n) + D_g(a_n, x_n) \leq D_g(x^*, x_n). \tag{34}$$

Since $u_n = P_C^g a_n$, from Lemma 1.3, we get:

$$D_g(x^*, u_n) + D_g(u_n, a_n) \leq D_g(x^*, a_n). \tag{35}$$

Substituting (35) into (34), we obtain:

$$D_g(x^*, \ u_n)+D_g(u_n, \ a_n)+D_g(a_n, \ x_n)\leq D_g(x^*, \ x_n), \tag{36}$$

which implies that

$$D_g(x^*, \ u_n)\leq D_g(x^*, \ x_n)-D_g(u_n, \ a_n)-D_g(a_n, \ x_n). \tag{37}$$

Using the same techniques of proof of Theorem 3.2 pp. 64 of [18] from (19), (9), (10), and Lemma 1.5, we get:

$$D_g(x^*, \ r_n)\leq \eta_{n,1}D_g(x^*, \ x_n)+\eta_{n,2}D_g(x^*, \ \nabla g^*Tx_n)+\eta_{n,3}D_g(x^*, \ u_n) \tag{38}$$

$$-\eta_{n,1}\eta_{n,2}\rho_r^*\left(\|\nabla gx_n - Tx_n\|\right). \tag{39}$$

From inequalities (37) and (39) above and assumption on T, we get:

$$D_g(x^*, \ r_n)\leq \eta_{n,1}D_g(x^*, \ x_n)+\eta_{n,2}D_g(x^*, \ x_n)+\eta_{n,3}D_g(x^*, \ u_n) \tag{40}$$

$$\leq D_g(x^*, \ x_n)-\eta_{n,3}\left[D_g(u_n, \ a_n)+D_g(a_n, \ x_n)\right] \tag{41}$$

$$-\eta_{n,1}\eta_{n,2}\rho_r^*\left(\|\nabla gx_n - Tx_n\|\right) \tag{42}$$

$$\leq D_g(x^*, \ x_n). \tag{43}$$

Now, from (19), Lemma 1.3 (ii), Lemma 1.4, and (44), we obtain:

$$D_g(x^*, \ x_{n+1})\leq D_g\left(x^*, \ \nabla g^*\left(\alpha_n\nabla gv+(1-\alpha_n)\nabla gr_n\right)\right) \tag{44}$$

$$\leq \alpha_nD_g(x^*, \ v)+(1-\alpha_n)D_g(x^*, \ x_n) \tag{45}$$

$$\leq \max\left\{D_g(x^*, \ v), D_g(x^*, \ x_n)\right\}, \tag{46}$$

and by induction, we get:

$$D_g(x^*, \ x_n)\leq \max\left\{D_g(x^*, \ v), D_g(x^*, \ x_0)\right\}. \tag{47}$$

Hence, the sequence $\left\{D_g(x^*, \ x_n)\right\}$ is bounded. Thus, by Lemma 7 in ref. [31], the sequence $\{x_n\}$ is bounded and so are $\{a_n\}$, $\{u_n\}$, $\{r_n\}$, $\{Gp_n\}$, and $\{Tx_n\}$.

Theorem 1.11 Suppose the Assumptions (A1) and (A2) hold. Let $G: C\to E^*$ and $T: C\to E^*$ be continuous pseudomonotone and BRGN mappings, respectively with $\Omega = VI(C, G)\cap F_g(T)\neq \varnothing$. Then, the sequens $\{x_n\}$ generated by Algorithm 1 converge strongly to an element $x^* = P_\Omega^g(v)$.

Proof: From Theorem 1.10 above, we know that the sequence $\{x_n\}$ is bounded. Let $x^* = P_\Omega^g(v)$. Now, using the same techniques of proof of Theorem 2 of ref. [32], we get:

$$D_g(x^*, \ x_{n+1})\leq (1-\alpha_n)D_g(x^*, \ x_n)+\alpha_n\|\nabla gv-\nabla gx^*\|\|x_n-w_n\| \tag{48}$$

$$+\alpha_n\langle\nabla gv-\nabla gx^*, x_n-x^*\rangle. \tag{49}$$

Furthermore, from (19), Lemma 1.3 (ii), and Lemma 1.4, we have:

$$D_g(x^*, \ x_{n+1})\leq D_g\left(x^*, \ \nabla g^*\left(\alpha_n\nabla gv+(1-\alpha_n) \ \nabla gr_n\right)\right) \tag{50}$$

$$\leq \alpha_n \ D_g(x^*, \ v)+(1-\alpha_n) \ D_g(x^*, \ r_n). \tag{51}$$

DOI: http://dx.doi.org/10.5772/intechopen.106547

Thus, from (51) and (43), we get:

$$D_g(x^*, \ x_{n+1}) \le \alpha_n \ D_g(x^*, \ v) + (1-\alpha_n) \ D_g(x^*, \ x_n) \tag{52}$$

$$-(1-\alpha_n) \ \eta_{n,3}\left[D_g(u_n, \ a_n) + D_g(a_n, \ x_n)\right] \tag{53}$$

$$-(1-\alpha_n)\eta_{n,1}\eta_{n,2}\rho_r^*\left(\|\nabla \ gx_n - Tx_n\|\right), \tag{54}$$

Now, to complete the proof we use the following two cases:

Case 1. Assume that there exists $n_0 \in \mathbb{N}$ such that the sequence $D_g(x^*, \ x_n)$ is decreasing for all $n \ge n_0$. It then follows that the sequence $D_g(x^*, \ x_n)$ converges, and hence, $D_g(x^*, \ x_n) - D_g(x^*, \ x_{n+1}) \to 0$ as $n \to \infty$. Thus, from (53), we obtain:

$$\lim_{n\to\infty}\left[D_g(u_n, \ a_n) + D_g(a_n, \ x_n)\right] = 0, \tag{55}$$

and

$$\lim_{n\to\infty}\rho_r^*\left(\|Tx_n - \nabla \ gx_n\| = 0.\right. \tag{56}$$

Hence, from (55) and Lemma 2.4 of [33] p. 15, we get:

$$\lim_{n\to\infty}\|u_n - a_n\| = \lim_{n\to\infty}\|x_n - a_n\| = 0. \tag{57}$$

From (56) and property of ρ_r^*, we obtain:

$$\lim_{n\to\infty}\|Tx_n - \nabla \ gx_n\| = 0. \tag{58}$$

From (58) and the fact that $\nabla \ g^*$ is uniformly continuous on bounded subsets of E^*, we obtain:

$$\lim_{n\to\infty}\|\nabla \ g^* Tx_n - x_n\| = 0. \tag{59}$$

Moreover, from (19) and Lemma 1.4, we get:

$$D_g(x_n, \ w_n) = D_g\left(x_n, \ \nabla \ g^*\left[\alpha_n \ \nabla \ gv + (1-\alpha_n) \ \nabla \ gr_n\right]\right) \tag{60}$$

$$\le \alpha_n \ D_g(x_n, \ v) + (1-\alpha_n) \ D_g(x_n, \ r_n) \tag{61}$$

$$= \alpha_n \ D_g(x_n, \ v) + (1-\alpha_n) \ \left[\eta_{n,1}D_g(x_n, \ x_n) + \eta_{n,2}D_g(x_n, \ \nabla \ g^* Tx_n)\right] \tag{62}$$

$$+(1-\alpha_n) \ \left[\eta_{n,3}D_g(x_n, \ u_n) + \eta_{n,4}D_g(x_n, \ v_n)\right] \tag{63}$$

Thus, from Lemma 2.4 of [33] p. 15, (57), (59), and (61), we get:

$$\lim_{n\to\infty} D_G(x_n, \ w_n) = 0, \tag{64}$$

which implies that

$$\lim_{n\to\infty}\|x_n - w_n\| = 0. \tag{65}$$

Now, since $\{x_n\}$ is bounded in C there exists $u \in C$ and a subsequence $\{x_{n_k}\}$ such that $\{x_{n_k}\}$ converges weakly to u and

$$\limsup_{n\to\infty}\langle x_n - x^*, \nabla gv - \nabla gx^*\rangle = \lim_{k\to\infty}\langle x_{n_k} - x^*, \nabla gv - \nabla gx^*\rangle. \tag{66}$$

From (58) and definition of T, we have $u \in F_g(T)$.
Next, we prove that $u \in VI(C, G)$. Since $a_n \in P_n$ then we can get:

$$0 = \left\langle Gp_{n_k}, a_{n_k} - p_{n_k}\right\rangle \tag{67}$$

$$= \left\langle Gp_{n_k}, a_{n_k} - x_{n_k}\right\rangle + \left\langle Gp_{n_k}, x_{n_k} - p_{n_k}\right\rangle \tag{68}$$

which implies that

$$\left\langle Gp_{n_k}, x_{n_k} - p_{n_k}\right\rangle = \left\langle Gp_{n_k}, x_{n_k} - a_{n_k}\right\rangle \le \|Gp_{n_k}\|\|x_{n_k} - a_{n_k}\|. \tag{69}$$

From (57), (69) and the fact that the sequence $\{Gp_n\}$ is bounded, we get:

$$\lim_{k\to\infty}\left\langle Gp_{n_k}, x_{n_k} - p_{n_k}\right\rangle = 0. \tag{70}$$

Now, we prove

$$\lim_{k\to\infty}\|P_C^g y_{n_k} - x_{n_k}\| = 0. \tag{71}$$

From (70), Lemma 1.9 and Lemma 2.4 of [33] p. 15, we get:

$$\lim_{k\to\infty}\tau_{n_k}\|P_C^g y_{n_k} - x_{n_k}\| = 0. \tag{72}$$

First, consider the case when $\liminf_{k\to\infty}\tau_{n_k} > 0$. In this case, there is a constant $\tau > 0$ such that $\tau_{n_k} \ge \tau > 0$ for all $k \in \mathbb{N}$. Thus, we have:

$$\|P_C^g y_{n_k} - x_{n_k}\| = \frac{1}{\tau_{n_k}}\tau_{n_k}\|P_C^g y_{n_k} - x_{n_k}\| \le \frac{1}{\tau}\tau_{n_k}\|P_C^g y_{n_k} - x_{n_k}\|. \tag{73}$$

Thus, from (72) and (73), we obtain:

$$\lim_{k\to\infty}\|P_C^g y_{n_k} - x_{n_k}\| = 0. \tag{74}$$

Second, we consider the case when $\liminf_{k\to\infty}\tau_{n_k} = 0$. In this case, we take a subsequence $\{n_{k_j}\}$ of $\{n_k\}$, if necessary, we assume without loss of generality that

$$\lim_{k\to\infty}\tau_{n_k} = 0 \ \ and \ \ \lim_{k\to\infty}\|x_{n_k} - P_C^g y_{n_k}\| = a > 0. \tag{75}$$

Consider $p'_{n_k} = \frac{1}{l}\tau_{n_k}P_C^g y_{n_k} + (1 - \frac{1}{l}\tau_{n_k})x_{n_k}$. Then, from (75), we have:

$$\lim_{k\to\infty} \|x_{n_k} - p'_{n_k}\| = \lim_{k\to\infty} \frac{1}{l}\tau_{n_k}\|x_{n_k} - P^g_C y_{n_k}\| = 0. \tag{76}$$

From the search rule in Step 2 and the definition of p'_{n_k}, we get:

$$\left\langle Gx_{n_k} - Gp'_{n_k}, x_{n_k} - P^g_C y_{n_k}\right\rangle > \mu D_g\left(P^g_C y_{n_k}, x_{n_k}\right). \tag{77}$$

Using (76), (77), and Lemma 2.4 of [33] p. 15, and the fact that G is uniformly continuous on bounded subsets of C, we obtain:

$$\lim_{k\to\infty} \|P^g_C y_{n_k} - x_{n_k}\| = 0,$$

which is a contradiction to (75). Therefore, the equality in (71) holds. Combining Lemma 1.3 and 17, we get:

$$\left\langle Gx_{n_k}, z - P^g_C z_{n_k}\right\rangle \geq \left\langle \nabla\ gx_{n_k} - \nabla\ gP^g_C y_{n_k}, z - P^g_C y_{n_k}\right\rangle, \forall z \in C, \tag{78}$$

which implies that

$$\left\langle Gx_{n_k}, y - x_{n_k}\right\rangle \geq \left\langle Gx_{n_k}, P^g_C y_{n_k} - x_{n_k}\right\rangle \tag{79}$$

$$+\left\langle \nabla\ gx_{n_k} - \nabla\ gP^g_C y_{n_k}, z - P^g_C y_{n_k}\right\rangle, \forall z \in C. \tag{80}$$

Thus, from (80), (78) and the fact that $\nabla\ g$ is uniformly continuous, we obtain:

$$\liminf_{k\to\infty} \left\langle Gx_{n_k}, z - x_{n_k}\right\rangle \geq 0, \forall z \in C. \tag{81}$$

Moreover, let $\{\xi_k\}$ be a sequence of decreasing numbers such that $\{\xi_k\} \to 0$ as $k \to \infty$ and w be an arbitrary element of C. Using inequality (81), we can find a large enough N_k such that

$$\left\langle Gx_{n_k}, w - x_{n_k}\right\rangle + \xi_k \geq 0, \forall k \geq N_k. \tag{82}$$

From (82) and the fact that $Gx_{n_k} \neq 0$, we get:

$$\left\langle Gx_{n_k}, \xi_k d_k + w - x_{n_k}\right\rangle \geq 0, \forall k \geq N_k, \tag{83}$$

for some $d_k \in C$ satisfying $\left\langle Gx_{n_k}, d_k\right\rangle = 1$. In addition, from the definition of G and inequality (83), we have:

$$\left\langle G(w + \xi_n d_k w), w + \xi_k d_k w - x_{n_k}\right\rangle \geq 0, \forall k \geq N_k, \tag{84}$$

which implies that

$$\left\langle Gw, w - x_{n_k}\right\rangle \geq \left\langle Gw - G(w + \xi_k d_k w), w + \xi_k d_k w - x_{n_k}\right\rangle \tag{85}$$

$$-\xi_k\langle Gw, d_k\rangle, \forall k \geq N_k, \tag{86}$$

Since $\xi_k \to 0$ as $k \to \infty$ and G is continuous, then from inequality (86), we obtain:

$$\langle Gw, w - u\rangle = \liminf_{k\to\infty} \langle Gw, w - x_{n_k}\rangle \geq 0, \forall w \in C. \tag{87}$$

Thus, $u \in VI(C, G)$, and hence, $u \in \Omega$. It follows Lemma 1.3 (i), that

$$\limsup_{n\to\infty}\langle x_n - x^*, \nabla gv - \nabla gx^*\rangle = \lim_{k\to\infty}\langle x_{n_k} - x^*, \nabla gv - \nabla gx^*\rangle \tag{88}$$

$$= \langle u - x^*, \nabla gv - \nabla gx^*\rangle \leq 0. \tag{89}$$

Therefore, from (49), (65), (89), and Lemma 2.5 of [34] p. 243, we conclude that $D_g(x^*, x_n) \to 0$ as $n \to \infty$. Hence, by Lemma 2.4 of [33] p. 15, $x_n \to x^*$ as $n \to \infty$.

Case 2. Suppose that there exists a subsequence $\{n_i\}$ of $\{n\}$ such that

$$D_g(x^*, x_{n_i}) < D_g(x^*, x_{n_i+1}), \forall i \in \mathbb{N}. \tag{90}$$

Then, by Lemma 3.1 of [35] p. 904, there exists a nondecreasing sequence $\{m_k\}$ in the set of natural numbers such that $m_k \to \infty$ as $k \to \infty$, $D_g(x^*, x_{m_k}) \leq D_g(x^*, x_{m_k+1})$ and $D_g(x^*, x_k) \leq D_g(x^*, x_{m_k+1})$ for all k elements of the set of natural numbers. Thus, from (53), we obtain:

$$\lim_{k\to\infty}\|u_{m_k} - a_{m_k}\| = \lim_{k\to\infty}\|Tx_{m_k} - \nabla gx_{m_k}\| = 0. \tag{91}$$

Moreover, following the methods in Case 1 above, we get:

$$\lim_{k\to\infty}\|x_{m_k} - w_{m_k}\|, \tag{92}$$

and

$$\limsup_{k\to\infty}\langle\langle x_{m_k} - x^*, \nabla gv - \nabla gx^*\rangle \leq 0. \tag{93}$$

In addition, from (49) and inequality (90) above, we obtain:

$$D_g(x^*, x_{m_k}) \leq \|x_{m_k} - r_{m_k}\|\nabla gv - \nabla gx^*\| \tag{94}$$

$$+\langle x_{m_k} - x^*, \nabla gv - \nabla gx^*\rangle. \tag{95}$$

Therefore, from (92), (93), and (95), we obtain $\lim_{k\to\infty} D_g(x^*, x_{m_k}) = 0$. But from inequality (53), we obtain that $\lim_{k\to\infty} D_g(x^*, x_{m_k+1}) = 0$, which implies that $\lim_{k\to\infty} D_g(x^*, x_k) = 0$. Thus, by Lemma 2.4 of [33] p. 15 $x_k \to x^*$ as $k \to \infty$.

We remark that the proof of Theorem 11 provides the following result for a common point in the solution set of VIP and the set of g−fixed point of continuous monotone and BRGN, mappings, respectively.

Theorem 1.12 Suppose the Assumptions (A1) and (A2) hold. Let $G : C \to E^*$ and $T : C \to E^*$ be continuous monotone and BRGN mappings, respectively with $\Omega = VI(C, G) \cap F_g(T) \neq \emptyset$. Then, the sequens $\{x_n\}$ generated by Algorithm 1 converge strongly to an element $x^* = P^g_\Omega(v)$.

If in Algorithm 1, we put $C = E$, then P_C^g is reduced to the identity mapping in E and $VI(C, G) = G^{-1}(0)$. Thus, we get the following Algorithm 2 for a common point in the set of zeros and the set of $g-$fixed point of continuous pseudomonotone and BRGN mappings, respectively.

Algorithm 2: For any $x_0, v \in E$, define an algorithm by.

Step 1. Compute

$$y_n = \nabla g^* [\nabla gx_n - \beta Gx_n] \text{ and } d(y_n) = x_n - y_n. \tag{96}$$

If $d(y_n) = 0$ and $\nabla gx_n - Tx_n = 0$, then stop and $x_n \in \Omega$. Otherwise,
Step 2. Compute $p_n = x_n - \tau_n d(y_n)$,
where $\tau_n = l^{j_n}$ and j_n is the smallest nonnegative integer j satisfying

$$\langle Gx_n - Gp_n, \ d(y_n) \rangle \le \mu D_g(y_n, \ x_n). \tag{97}$$

Step 3. Compute

$$\begin{cases} u_n = P^f_{P_n} \nabla g^* (\nabla gx_n - \beta Gp_n), \\ r_n = \nabla g^* (\eta_{n,1} \nabla gx_n + \eta_{n,2} Tx_n + \eta_{n,3} \nabla gu_n), \\ x_{n+1} = \nabla g^* (\alpha_n \nabla gv + (1 - \alpha_n) \nabla gr_n), \end{cases} \tag{98}$$

where $g \in \mathcal{G}(E)$, $P_n = \{p \in C : \langle Gp_n, p - p_n \rangle = 0\}$, and $\{\eta_{n,i}\} \subset [\epsilon, 1) \subset (0, 1)$, for $i = 1,2,3$ such that $\sum_{i=1}^{3} \eta_{n,i} = 1$, $\forall n \ge 0$.
Step 4. Set $n := n + 1$ and go to **Step 1**.

Corollary 1.13 Suppose the Assumptions (A1) and (A2) hold. Let $G : E \to E^*$ and $T : E \to E^*$ be continuous pseudomonotone and BRGN mappings, respectively with $\Omega = G^{-1}(0) \cap F_g(T) \ne \emptyset$. Then, the sequens $\{x_n\}$ generated by Algorithm 2 converge strongly to an element $x^* = P_\Omega^g(v)$.

If in Algorithm 2, we put $T = \nabla g$, the identity mapping in E, then we get the following corollary for zero point of continuous pseudomonotone.

Corollary 1.14 Suppose the Assumptions (A1) and (A2) hold. Let $G : E \to E^*$ be a continuous pseudomonotone mapping with $G^{-1}(0) \ne \emptyset$. Then, the sequens $\{x_n\}$ generated by Algorithm 2 converge strongly to an element $x^* = P^g_{G^{-1}(0)}(v)$.

2.1 Application to convex minimization problem

In this section, we apply Corollary 1.14 to find the minimum point of the convex function in Banach Spaces.

Let $f : E \to \mathbb{R}$ be a convex smooth function. We consider the problem of finding a point $z \in E$ such that

$$f(z) = \min_{x \in E} \{f(x)\}. \tag{99}$$

According to Fermat's rule, this problem is equivalent to the problem of finding $z \in E$ such that

$$\nabla fz = 0, \tag{100}$$

where ∇f is a gradient of f. We note that ∇f is monotone mapping (see, e.g., [36, 37]) and hence pseudomonotone mapping.

Now, if in Algorithm 2, we assume $G = \nabla f$, then we obtain the following Algorithm 3 for the minimum point problem of convex functions in real reflexive Banach spaces.

Algorithm 3: For any $x_0, v \in E$, define an algorithm by.

Step 1. Compute

$$y_n = \nabla g^* \left[\nabla gx_n - \beta \nabla fx_n\right] \text{ and } d(y_n) = x_n - y_n. \tag{101}$$

If $d(y_n) = 0$, then stop and $x_n \in \Omega$. Otherwise,

Step 2. Compute $p_n = x_n - \tau_n d(y_n)$,

where $\tau_n = l^{j_n}$ and j_n is the smallest nonnegative integer j satisfying

$$\langle \nabla fx_n - \nabla fp_n, \ d(y_n)\rangle \le \mu D_g(y_n, \ x_n). \tag{102}$$

Step 3. Compute

$$\begin{cases} u_n = P^f_{P_n} \nabla g^* (\nabla gx_n - \beta \nabla fp_n), \\ r_n = \nabla g^* (\eta_n \nabla gx_n + (1-\eta_n)\nabla gu_n), \\ x_{n+1} = \nabla g^* (\alpha_n \nabla gv + (1-\alpha_n)\nabla gr_n), \end{cases} \tag{103}$$

where $g \in \mathcal{G}(E)$, $P_n = \{p \in C : \langle \nabla fp_n, \ p - p_n\rangle = 0\}$, and $\{\eta_n \subset [\epsilon, 1) \subset (0, 1), \forall n \ge 0$.

Step 4. Set $n := n + 1$ and go to **Step 1**.

The method of proof Theorem 1.11 provides the proof of the following theorem of finding the minimum point of a convex function in reflexive Banach spaces.

Theorem 1.15 Suppose the Assumptions (A1) and (A2) hold. Let $f : E \to \mathbb{R}$ be a convex smooth function with ∇f is continuous and $\Omega = \{z : f(z) = \min_{x \in E} f(x)\} \ne \emptyset$. Then, the sequens $\{x_n\}$ generated by Algorithm 3 converge strongly to an element $x^* = P^g_\Omega(v)$.

2.2 Numerical example

In this section, we provide a numerical example to explain the conclusion of our main result. The following example verifies the conclusion of Theorem 1.11.

Example 1.16. Let $E = \mathbb{R}$ be with the standard topology. Define $g : \mathbb{R} \to \mathbb{R}$, by $g(x) = \frac{x^2}{2}$, then $g^*(x^*) = \frac{x^{*2}}{2}$ and $\nabla g(x) = x = \nabla g^*(x^*) = x^*$, where $x = (x_1, x_2, x_3) \in \mathbb{R}$. Let $C = \{x \in \mathbb{R} : \|x\| \le 1\}$. Let $G, T : C \to \mathbb{R}$ be defined by $G(x_1, x_2, x_3) = (x_1, x_2, x_3)\left(1.8 - \sqrt{x_1^2 + x_2^2 + x_3^2}\right)$ and $T(x_1, x_2, x_3) = \frac{(x_1, x_2, x_3)}{5}$, then G is continuous pseudomonotone mapping and T is BRGN mapping with $\Omega = VI(C, G) = \{0\} = F_g(T) \ne \emptyset$. Now, if we assume $v = (v_1, v_2, v_3) = (0, 0.5, 0.5)$, $\alpha_n = \frac{1}{n+10}$, $\eta_{n,1} = \eta_{n,2} = 0.001 + \frac{1}{n+1000}$ and $\eta_{n,3} = 0.998 - \frac{2}{n+1000}$, $l = 0.8$, $\mu = 0.9$ and $\lambda = 1$ for all $n \ge 0$, and take different initial points $x_0 = (0, 1, -1)$, $x_0' = (1.2233, 2, -1.4532)$

n	x_n	x_n	x_n
0	$(0.0000, 1.0000, -1.0000)$	$(1.2233, 2.0000, -1.4532)$	$(1.0000, 2.0000, 3.0000)$
1	$(0.0000, 0.2379, 0.1881)$	$(-0.1729, 0.1325, 0.9238)$	$(0.0195, 0.0622, -0.8500)$
10	$(0.0000, 0.0335, 0.0228)$	$(-0.0868, 0.0280, 0.0797)$	$(0.0110, 0.0269, -0.2428)$
100	$(0.0000, 0.0058, 0.0040)$	$(-0.0134, 0.0049, 0.0026)$	$(0.0019, 0.0047, 0.0024)$
200	$(0.0000, 0.0030, 0.0021)$	$(-0.0069, 0.0025, 0.0014)$	$(9.8943e^{-04}, 0.0024, 0.0012)$
300	$(0.0000, 0.0020, 0.0014)$	$(-0.0046, 0.0017, 9.1348e^{-04})$	$(6.6775e^{-04}, 0.0017, 8.4377e^{-04})$
400	$(0.0000, 0.0016, 0.0011)$	$(-0.0035, 0.0013, 6.9030e^{-04})$	$(5.0352e^{-04}, 0.0013, 6.3737e^{-04})$
500	$(0.0000, 0.0012, 8.5023e^{-04})$	$(-0.0028, 0.0010, 5.5401e^{-04})$	$(4.0389e^{-04}, 0.0010, 5.1207e^{-04})$
	$\vdots$	$\vdots$	$\vdots$
	$\downarrow$	$\downarrow$	$\downarrow$
	$(0,0,0)$	$(0,0,0)$	$(0,0,0)$
The sequence $\{x_n\}$, generated by Algorithm 1 converges strongly to $x^* = (0, 0, 0)$.			

Table 1.
Convergence of the sequence $\{x_n\}$ generated by Algorithm 1 for different choices of x_0.

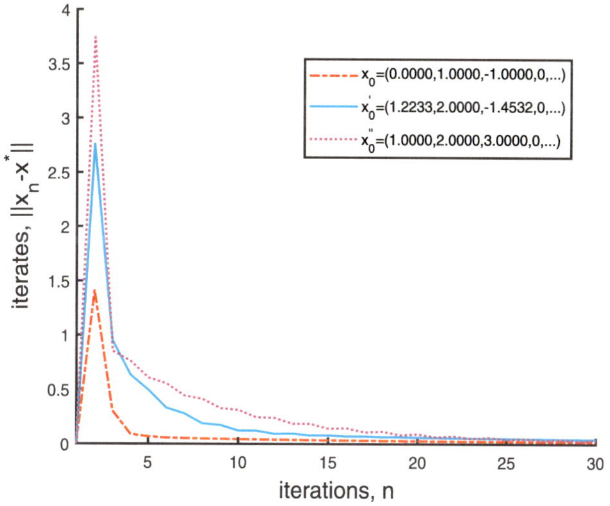

Figure 1.
The graph of $\|x_n - x^*\|$ versus number of iterations with different choices of x_0.

and $x_0'' = (1,2,3)$, then in all cases the numerical example result using MATLAB provides that the sequence $\{x_n\}$, generated by Algorithm 1 converges strongly to $x^* = (0,0,0)$ (see, **Table 1**). In addition, we have sketched the error term $\|x_n - x^*\|$ for each initial point. From the sketch, we observe that $\|x_n - x^*\| \to 0$ as $n \to \infty$ (see, **Figure 1**).

3. Conclusions

In this, manuscript, we introduced an iterative method for approximating a common solution of VIP of continuous pseudomonotone and GFPP of BRGN mappings and proved strong convergence of the sequence generated by the method to a common solution in real reflexive Banach spaces. In addition, we gave an application of our main result to find a minimum point of convex functions in real reflexive Banach spaces. Finally, a numerical example that supports our main result is presented. Our results extend and generalize many results in the literature. In particular, Theorem 1.11 extends the results in [3, 4, 7, 13, 16, 17, 38] from real Hilbert spaces to real reflexive Banach spaces. Moreover, Theorem 1.11 extends the classes of mappings in Theorem 3.1 of Tufa and Zegeye [17] and Theorem 3.2 of Wega and Zegeye [18] from Lipschitz monotone mapping to continuous pseudomonotone mappings in reflexive real Banach spaces.

Conflict of interest

The authors declare that they have no competing interests.

Author details

Getahun Bekele Wega
Jimma University, Jimma, Ethiopia

*Address all correspondence to: getahunbekele2012@gmail.com

References

[1] Hartman P, Stampacchia G. On some non-linear elliptic differential-functional equations. Acta Mathematica. 1966;**115**: 271-310

[2] Cai G, Bu S. An iterative algorithm for a general system of variational inequalities and fixed point problems in q-uniformly smooth Banach spaces. Optimization Letters. 2013;7:267-287

[3] Iiduka H, Akahashi W, Toyoda M. Approximation of solutions of variational inequalities for monotone mappings. Mathematical Journal. 2004; **14**:49-61

[4] Nadezhkina N, Takahashi W. Weak convergence theorem by an extragradient method for nonexpansive mappings and monotone mappings. Journal of Optimization Theory and Applications. 2006;**128**: 191-201

[5] Noor M. A class of new iterative methods for solving mixed variational inequalities. Mathematical and Computer Modelling. 2000;**31**:11-19

[6] Thong DV, Hieu D. Weak and strong convergence theorems for variational inequality problems. Numerical Algorithms. 2018;**78**:1045-1060

[7] Yao Y, Postolache M, Liou Y-C. Extragradient-type method for monotone variational inequalities. Fixed Point Theory Applications. 2013;**2013**: 1-15

[8] Zegeye H, Ofoedu E, Shahzad N. Convergence theorems for equilibrium problem, variotional inequality problem and countably infinite relatively quasi-nonexpansive mappings. Applied Mathematics and Computation. 2010; **216**:3439-3449

[9] Zegeye H, Shahzad N. A hybrid scheme for finite families of equilibrium, varational inequality and fixed point problems. Nonlinear Analysis. 2011;**74**: 263-272

[10] Zegeye H, Shahzad N. Construction of a common solution of a finite family of variational inequality problems for monotone mappings. Nonlinear Science Applications. 2016;**9**:1645-1657

[11] Kumam P. A hybrid approximation method for equilibrium and fixed point problems for a monotone mapping and a nonexpansive. Nonlinear Analysis: Hybrid Systems. 2008;**2**:1245-1255

[12] Reich S, Sabach S. Three strong convergence theorems regarding iterative methods for solving equilibrium problems in reflexive Banach spaces. Contemporary Mathematics. 2012;**568**: 225-240

[13] Takahashi W, Toyoda M. Weak convergence theorems for nonexpansive mappings and monotone mappings. Journal of Optimization Theory and Applications. 2003;**118**:417-428

[14] Takahashi W, Zembayashi K. Strong and weak convergence theorems for equilibrium problems and relatively nonexpansive mappings in Banach spaces. Nonlinear Analysis. 2009;**70**: 45-57

[15] Zegeye H, Shahzad N. Strong convergence for monotone mappings and relatively weak nonexpansive mappings. Nonlinear Analysis. 2009;**70**: 2707-2716

[16] Zhanga Y, Yuanb Q. Iterative common solutions of fixed point and variarional inequality problems. Journal

of Nonlinear Science and Applications. 2016;**9**:1882-1890

[17] Tufa A, Zegeye H. An algorithm for finding a common point of the solutions of fixed point and variational inequality problems in Banach spaces. Arabian Journal of Mathematics. 2015;**4**:199-213

[18] Wega G, Zegeye H. Convergence theorem of common solution of variational inequality and fixed point problems in Banach spaces. Applied Set-Valued Analysis and Optimization. 2021; **3**:55-73

[19] Bonnans J, Shapiro A. Perturbation Analysis of Optimization Problem. New York: Springer; 2000

[20] Bauschke HH, Borwein J. M, Legendre functions and the method of random Bregman projections. Journal of Convex Analysis. 1997;**4**:27-67

[21] Censor Y, Lent A. An iterative row-action method for interval convex programing. Journal of Optimization Theory and Applications. 1981;**34**(3): 321-353

[22] Butnariu D, Iusem A. N, Totally Convex Functions for Fixed Points Computation and Infinite Dimentional Optimization. Vol. 40. Dodrecht, The Netherlands: Klumer Academic; 2000

[23] Reich S, Sabach S. Two strong convergence theorem for Bregman strongly nonexpansive operators in reflexive Banach spaces. TMA. 2010;**73**: 122-135

[24] Reich S, Sabach S. A projection method for solving nonlinear problems in reflexive Banach spaces. Journal of Fixed Point Theory and Applications. 2011:101-116

[25] Butnariu D, Resmerita E. Bregman distances, totally convex functions and a method for solving operator equations in Banach spaces. Abstract Application Analysis. 2006;**2006**:1-39

[26] Senakka P, Cholamjiak P. Approximation method for solving fixed point problem of Bregman strongly nonexpansive mappings in reflexive Banach spaces. Elementary Education. 2016;**65**(1):209-220

[27] Phelps RP. Convex Functions, Monotone Operators, and Differeniability. 2nd ed. Vol. 1364. Berlin: Springer Verlag; 1993

[28] Zalinescu C. Convex Analysis in General Vector Spaces. River Edge, NJ, USA: World Scientific; 2002

[29] Nesterov Y. Introductory Lectures on Convex Optimization: A Basic Course. Kluwer Academic Publishers; 2004. pp. 63-64

[30] Naraghirad E, Yao CJ. Weak relatively nonexpanseve mappings in Banach spaces. Fixed Point Theory and Applications. 2013;**2013**:141

[31] Wega G, Zegeye H. Convergence results of forward-backward method for a zero of the sum of maximally monotone mappings in Banach spaces. Computational and Applied Mathematics. 2020;**39**:223. DOI: 10.1007/s40314-020-01246-z

[32] Wega G, Zegeye H. Approximation of a common f-fixed point of f-pseudocontractive mappings in Banach spaces. Rendiconti del Circolo Matematico di Palermo Series. 2020;**270**: 1139-1162

[33] Naraghirad E, Yao JC. Bergman weak relatively nonexpansive mappings

in Banach spaces. Fixed Point Applications. 2013;**2013**:141

[34] Xu H-H. Viscosity approximation methods for nonexpansive mappings. Journal of Mathematical Analysis and Applications. 2004;**298**(1):279-291

[35] Maingé P. Strong convergence of projected subgradiant method for nonsmooth and nonstrictily convex minimization. Set-Valued Analysis. 2008;**16**:899-912

[36] Baillon J, Haddad G. Quelques proprietes des operateurs angle-bornes et cycliquement monotones. Israel Journal of Mathematics. 1977;**26**:137-150

[37] Rockafellar R. On the maximal monotonicity of subdifferential mappings. Pacific Journal of Mathematics. 1970;**33**:209-216

[38] Tada A, Takahashi W. Weak and strong convergence theorems for nonexpansive mappings and equilbrium problems. Journal of Optimization Theory and Applications. 2007;**133**: 359-370

Section 3

Stability and Derivatives for Fixed Points

Chapter 5

On Fixed Point for Derivative of Set-Valued Functions

Mohamad Muslikh and Adem Kilicman

Abstract

In this article, we showed the existence of a fixed point for the derivative of interval-valued functions. The investigation of the existence of such fixed points utilizes the common fixed point concepts for two mappings. Under the condition of compatibility of the hybrid composite mappings in the sense of the Pompei-Hausdorff metric the existence of a fixed point for the derivative is shown. Some examples to support the usability of the result of this study are also given.

Keywords: common fixed point theorem, set-valued maps, compatible mappings, differentiable maps, interval-valued functions

1. Introduction

E. Dyer in [1] conjectured that f and g must have a common fixed point in $[0, 1]$ if $f(g(t)) = g(f(t))$ for each $t \in [0, 1]$. In 1967, W.M. Boyce [2] replied in his paper that Dyer's question is negative as well as an answer from Husein [3] and Singh [4]. However, many researchers are curious about conjecture. In 1976, G Jungck [5] shows the existence of the common fixed point for two mappings by the commuting mapping method in general metric spaces. Since then the common fixed point research had quickly grown. In development, some of the researchers not only involved two mappings (single-valued mappings) but also they are more than it is [6]. In fact, some involve the set-valued mapping forms [7, 8].

In progress, the composition mappings are discussed not only between fellow of single-valued mappings or set-valued mappings but also its combination (mixed compositions between of single-valued and set-valued mappings). Since then several authors have studied common fixed point theorems for such mapping in different ways ([9–11] and references therein).

Itoh et al. [12] introduce "commute" term of hybrid composite functions in 1977. By this properties, they have proven common fixed point theorems in topological vector spaces. In 1982, Fisher [13] has introduced common fixed point theorems for commuting mappings in the sense of the other in metric spaces. Then Imdad [14] mentioned the properties $fFx \subseteq Ffx$ as "quasi-commute" to distinguish with the latter term. Whereas two commuting mappings F and f are weakly commuting, but in general two weakly commuting mappings do not commute as it is shown in Example 1 of [15].

In 1989, Kaneko [16] introduced the concept of "compatible" by using the Hausdorff metric and proved the existence of a common fixed point theorem by the concept. In 1993, Jungck [17] introduced the same things but used the concept of "δ-compatible" mappings in metric spaces and proved some common fixed point theorems for δ-compatible mappings.

Regarding the fixed point for derivatives has been observed by M. Elekes at all in [18]. In his paper, he shows the compositions of two functions derivatives have fixed points. This result is an affirmative answer to a question of K. Ciesielski, whether the composition of two derivatives on interval closed has a fixed point? The fixed point for a function is usual but for its derivatives is another something. Here have we the quadruplets (X, x, f, f'). How do these problems? By the device of commutativity and compatibility between the function and its derivatives, the author shows that the function derivatives of the real-valued function have a fixed point [19].

Motivated by the results mentioned above, in this article, we introduced the existence theorem of a fixed point for gh-derivative of the interval-valued function. To this work, we used hybrid composite mappings involving gh-derivative under the compatibility conditions.

2. gh-Differences

Suppose (X, d) is a metric spaces. The collection of all non-empty subsets of X is denoted by $\mathcal{P}_0(X)$. Whereas, the notation $\mathcal{B}(X)$ (resp. $\mathcal{CB}(X)$, $\mathcal{K}(X)$ and $\mathcal{KC}(X)$) is the collection of all non-empty bounded (resp. closed-bounded, compact and compact-convex) subsets of X.

In 1905, In his PhD thesis [20], Pompeiu defined the notions of *e′cart* between two sets. Hausdorff [21] studies the notion of set distance in the natural setting of metric spaces and with a small modification (the the sum is replaced by the maximum).

Let (X, d) be a metric spaces and $A, B \subset X$. The Hausdorff distance between A and B is a distance function $H : \mathcal{P}_0(X) \times \mathcal{P}_0(X) \to \mathbb{R}^+$ which is defined as

$$H(A, B) = \sup\{d(A,\ B), d(B, A)\}, \tag{1}$$

where $d(A, B) = \sup_{a \in A} d(a, B)$. Certainly value that $d(A, B) \neq d(B, A)$. The distance functions H to be a metric on the collection of all non-empty closed-bounded subset of X, $\mathcal{CB}(X)$. The metric spaces $(\mathcal{CB}(X), H)$ is called a complete metric spaces if the metric space X is a complete.

Suppose $\mathcal{I}(\mathbb{R}) = \{I = [a^-,\ a^+] | a^-, a^+ \in \mathbb{R}, a^- < a^+\}$. In [22], R.E. Moore et al. introduced an absolute value of the interval $J = [x^-, x^+]$ asfollows.

$$\|J\| = \max\{|x^-|, |x^+|\}. \tag{2}$$

For a given interval $I = [a^-, a^+]$ define the width, midpoint and radius of I, respectively, by

$$w(I) = a^+ - a^-, m(I) = \frac{1}{2}(a^- + a^+), and \quad r(I) = \frac{1}{2}(a^+ - a^-) \geq 0, \tag{3}$$

so that $a^- = m(I) - r(I)$ and $a^+ = m(I) + r(I)$. Thus the interval notasion $I = [a^-, a^+]$ can be written as the pair $I = (m(I); r(I))$..

The Pompeiu-Hausdorff distance on $\mathcal{I}(\mathbb{R})$ defined as

$$H(I, J) = \max\{|a^- - b^-|, |a^+ - b^+|\}, \tag{4}$$

where $I = [a^-, a^+]$ and $J = [b^-, b^+]$. The pair $(\mathcal{I}(\mathbb{R}), H)$ is a complete and separable metric space.

In 1967, M. Hukuhara [23] introduced the difference (h-difference) between U and V defined as $U \overset{h}{-} V = W$ if and only if $U = V + W$ for each $U,V,W \in \mathcal{KC}(\mathbb{R}^k)$. An important properties of the Hukuhara difference is that $U \overset{h}{-} U = \{\Theta\}$ and $(U + V) \overset{h}{-} V = U$. The Hukuhara difference is unique, but it does not always exists.

The Hukuhara difference had generalized by Markov in [24]. He defined is following as

$$U \overset{gh}{-} V = W \Leftrightarrow (\mathbf{a}) U = V + W \quad or \quad (\mathbf{b}) V = U + (-1)W. \tag{5}$$

Furthermore, Hukuhara difference generalized is called the gh-difference.

Both the equation $U = V + W$ and the equation $V = U + (-1)W$ can simultaneously holds. It is clear that h-difference is part of gh-difference. Therefore, the gh-difference is often said to be a generalization of the h difference. The gh difference of two intervals in $\mathcal{I}(\mathbb{R})$ always exists.

Proposition 1. Suppose $I = [a^-, a^+]$ and $J = [b^-, b^+]$ are intervals in $\mathcal{I}(\mathbb{R})$. The gh-difference of two intervals I and J always exists and

$$I \overset{gh}{-} J = [a^-, a^+] \overset{gh}{-} [b^-, b^+] = [c^-, c^+] \tag{6}$$

where $c^- = \min\{(a^- - b^-), (a^+ - b^+)\}$ and $c^+ = \max\{(a^- - b^-), (a^+ - b^+)\}$.

In [25, 26] defined $H(I, J) = \|I \overset{gh}{-} J\|$ for each $I,J \in I(\mathbb{R})$. An immediate property of the gh-difference for $I,J \in I(\mathbb{R}))$ is

$$H(I, I) = 0 \Leftrightarrow I \overset{gh}{-} J = 0 \Leftrightarrow I = J \tag{7}$$

It is also well known that $(I(\mathbb{R}), H)$is complete metric space.

2.1 gh-Derivative of set-valued functions

The mapping $F : X \to \mathcal{P}_0(Y)$ is called ***set-valued functions*** where the maps $F(x) \in \mathcal{P}_0(Y)$ for each $x \in X$. The function $f : X \to Y$ is said to be ***selection*** of F if $f(x) \in F(x)$ for all $x \in X$. We say that a point $z \in X$ is a fixed point of F if $z \in F(z)$.

The gh-derivative for an interval-valued function, expressed in terms of the difference quotient by gh-difference, has been first introduced in 1979 by S. Markov. A very recent and complete description of the algebraic properties of gh-derivative can be found in [27].

Definition 1 Let $F : [a, b] \to \mathcal{I}(\mathbb{R})$ be an interval-valued function and suppose t_0,$t_0 + h \in (a, b)$. The gh-derivative $F'_{gh}(t_0) \in \mathcal{I}(\mathbb{R})$ defined as

$$F'_{gh}(t_0) = \lim_{h \to 0} \frac{F(t_0 + h) \overset{gh}{-} F(t_0)}{h}. \tag{8}$$

If the limit, $\lim_{h \to 0} \frac{F(t_0+h) \overset{gh}{-} F(t_0)}{h}$ exists and satisfies Eq. (6), then F is said differentiable in the sense of generalized Hukuhara difference or gh-differentiable at a point $t_0 \in (a, b)$. The set-valued F'_{gh} is called a generalized Hukuhara derivative.

Theorem 1.1 If interval-valued functions $F : [a, b] \to \mathcal{I}(\mathbb{R})$ is a gh-differentiable at a point $p \in (a, b)$ then F is continuous at p.

Proof:

$$\begin{aligned} \lim_{x \to p} F(x) \overset{gh}{-} F(p) &= \lim_{x \to p} \left[\frac{F(x) \overset{gh}{-} F(p)}{(x-p)} (x-p) \right] \\ &= \left[\lim_{x \to p} \frac{F(x) \overset{gh}{-} F(p)}{(x-p)} \right] \left[\lim_{x \to p} (x-p) \right] \\ &= F'_{gh}(p) \cdot 0 = 0. \end{aligned}$$

So F is continuous at the point $p \in [a, b]$.

Theorem 1.2 [26] Let $F : [a, b] \to \mathcal{I}(\mathbb{R})$ be an interval-valued functions and $F(x) = [f(x), g(x)]$, where $f, g : [a, b] \to \mathbb{R}$. F is gh-differentiable on (a,b) if and only if f and g are differentiable on (a, b) and

$$F'_{gh}(x) = \left[\min \{f'(x),\ g'(x)\},\ \max \{f'(x),\ g'(x)\} \right],$$

for all $x \in (a, b)$.

This means that

$$F'_{gh}(x) = \begin{cases} [f'(x),\ g'(x)] & \text{if} f'(x) < g'(x), \\ [g'(x),\ f'(x)] & \text{if} f'(x) < g'(x) \end{cases}$$

for all $x \in (a, b)$.

3. Common fixed point

Definition 2 Suppose (X, d) is a metric space, $E \subset X$, $F : E \to \mathcal{B}(X)$ is a set-valued mapping and $f : E \to X$ is single-valued mapping.

i. F and f are said to **quasi commute** if $fFx \subseteq Ffx$ for each $x \in E$

ii. F and f are said to **commute** if $fFx = Ffx$ for each $x \in E$

iii. F and f are said to **slightly commute** if $fFx \in \mathcal{B}(X)$ for each $x \in E$ and $\delta(fFx, Ffx) \leq \max \{\delta(fx,\ Fx), diam(Fx)\}$

iv. F and f are said to **weakly commute** if $fFx \in \mathcal{B}(X)$ for each $x \in E$ and $\delta(fFx, Ffx) \leq \max \{\delta(fx,\ Fx), diam(fFx)\}$.

Suppose (X, d) is a metric space. The mapping $f,g : X \to X$ is a single-valued function or function and $F,G : X \to \mathcal{B}(X)$ is a set-valued function. For each $x,y \in X$, we used the notation as follows.

$$\mathcal{M}(F,f) = \max\{d(fx, Fx), d(fy, Fy), d(fx, Fy), d(fy, Fx), d(fx, fy)\}. \quad (9)$$

and

$$\mathcal{N}(F,f) = \max\left\{d(fx, fy), d(fx, Fx), d(fy, Fy), \frac{1}{2}[d(fx, Fy) + d(fy, Fx)]\right\}. \quad (10)$$

and

$$\mathcal{M}(F, G, f, g) = \max\{d(fx, gy), \delta(fx, Gy), \delta(gy, Fx)\}. \quad (11)$$

The following is the existence of common fixed point theorem that result by B. Fisher [13].

Theorem 1.3 Suppose (X, d) is a complete metric space, $F : X \to \mathcal{B}(X)$ is a set-valued mapping and $f : X \to X$ is a single-valued mapping satisfying the inequality

$$\delta(Fx, Fy) \le c\mathcal{M}(F,f) \quad (12)$$

for all $x,y \in X$, where $0 \le c < 1$. If.

A. f is continuous,

B. $F(X) \subseteq f(X)$, and

C. F and f are commute,

then F and f have a unique common fixed point.

B. Fisher also shown the same with assumes the continuity of F in X instead of the continuity of f [28].

The following theorem is generalization of Theorem 1.3 that has been resulted by M Imdad et al. [14].

Theorem 1.4 Suppose (X, d) is a complete metric space, $F : X \to \mathcal{B}(X)$ is a set-valued mapping and $f : X \to X$ is a single-valued mapping satisfying the inequality

$$\delta(Fx, Fy) \le \psi\mathcal{M}(F,f)$$

for all $x,y \in X$, where $\psi : [0, \infty) \to [0, \infty)$ is a nondecreasing, right continuous and $\psi(t) < t$, for all $t > 0$. If this following is satisfied

A. the function f is continuous,

B. the image of $F(X)$ is a subset of $f(X)$,

C. the set-valued F and single-valued f are weakly commute, and

D. $\exists\, x_0 \in X$ such that $\sup\{\delta(Fx_n, Fx_1) : n = 0, 1\cdots\} < +\infty$, then F and f have a unique common fixed point on X.

Theorem 1.5 Suppose (X, d) is a complete metric space, $F : X \to \mathcal{B}(X)$ is a set-valued mapping and $f : X \to X$ is a single-valued mapping satisfying the inequality

$$\delta(Fx, Fy) \leq \psi \mathcal{M}(F, f) \tag{13}$$

for all $x,y \in X$, where $\psi : [0, \infty) \to [0, \infty)$ is a non-decreasing, right continuous and $\psi(t) < t$, for all $t > 0$. If this following is satisfied

A. the set-valued mapping F or the single-valued mapping f are continuous,

B. the image $F(X)$ is a subset of the image $f(X)$,

C. the set-valued F and the singel valued f are slightly commute, and

D. $\exists\, x_0 \in X$ such that $\sup\{\delta(Fx_n, Fx_1) : n = 0, 1 \cdots\} < +\infty$,

then F and f have a unique common fixed point on X.

In the other context, Kaneko and Sessa in [16] introduce the “compatibility” term for the set-valued mapping F and the single-valued mapping f defined as follows:

Definition 3 Let (X, d) be a metric spaces. Suppose that $F : X \to \mathcal{CB}(X)$ is a set-valued mapping and $f : X \to X$ is a single-valued mapping. The mappings F and f is called **compatible** if the composition $fFx \in \mathcal{CB}(X)$ and the sequence $H(Ffx_n, fFx_n) \to 0$ whenever $\{x_n\}$ is sequence in X such that $fx_n \to t \in B \in \mathcal{CB}(X)$ and $Fx_n \to B \in \mathcal{CB}(X)$.

By using such the notion obtained the following theorem and lemma [16].

Theorem 1.6 Suppose (X, d) is a complete metric space, $F : X \to \mathcal{CB}(X)$ is a set-valued mapping, and $f : X \to X$ is a single-valued mapping satisfying the inequality

$$H(Fx, Fy) \leq c\mathcal{N}(F, f) \tag{14}$$

for all $x,y \in X$, where $0 \leq c < 1$. If this following is satisfied.

A. the set-valued mapping F and the single-valued mapping f are continuous,

B. the image $F(X)$ is a subset of the image $f(X)$, and

C. the set-valued F and the single-valued f are compatible,

then there exists a point $z \in X$ such that $f(z) \in F(z)$.

Lemma 1 Let (X, d) be a metric spaces. Suppose that $F : X \to \mathcal{CB}(X)$ and $f : X \to X$ are a compatible. If $fw \in Fw$ for some $w \in X$, then $Ffw = fFw$.

4. Fixed point for derivative

In this discussion, we shall make frequent use of the following Lemmas.

Lemma 2 [29] Let (X, d) be a metric spaces. If $C,D \in \mathcal{K}(X)$ and $c \in C$, then there exists the points $d \in D$ such that $d(c, d) \leq H(C, D)$.

Lemma 3 (Lemma 1 [4]) Let $\psi : [0, \infty) \to [0, \infty)$ be a real function such that non-decreasing, right continuous on $[0, \infty)$.

$$\lim_{n\to\infty} \psi^n(t) = 0.$$

if and only if for every $t > 0$ and $\psi(t) < t$.

In this result, we found that Lemma 1 also conversely holds provided its values of mapping are compact sets.

Lemma 4 Let (X, d) be a metric spaces and the set-valued $F : X \to \mathcal{K}(X)$ is a continuous on X. If there exists single-valued $f : X \to X$ is continuous on X such that $fw \in Fw$ for some $w \in X$, then the mappings F and f are compatible.

Proof: Since $Fx \in \mathcal{K}(X)$ for each $x \in X$ and f is continuous, the composition $fFx \in \mathcal{K}(X)$ for all $x \in X$. Suppose that the sequence $\{x_n\}$ on X such that the sequence of sets Fx_n converges to $K \in \mathcal{K}(X)$ and the sequence function fx_n converges to $z \in K$. In this case, we choose $z \in X$ such that $fz \in Fz$. Since F and f are continuous, we obtained

$$\begin{aligned}\lim_{n\to\infty} H(Ffx_n, fFx_n) &\le \lim_{n\to\infty} [H(Ffx_n, Fz) + H(Fz, \{fz\}) + H(\{fz\}, fFx_n)] \\ &= H(Fz, Fz) + H(Fz, \{fz\}) + H(\{fz\}, fK) \\ &= 0.\end{aligned}$$

The pairs F and f are proved as compatible by Definition 3.

By using the Lemma 4 we obtain the theorem as follows:

Theorem 1.7 Let (X, d) be a complete metric space, $F : X \to \mathcal{K}(X)$ be a continuous. Suppose there exists $f : X \to X$ is continuous on X such that $F(X) \subseteq f(X)$ and for all $x, y \in X$ satisfying the inequality

$$H(Fx, Fy) \le c\mathcal{N}(F, f), \tag{15}$$

where $0 \le c < 1$. Then $fz \in Fz$ for some $z \in X$ if and only if the pairs F and f are compatible.

Proof: Let $x_0 \in X$ be an arbitrary. Since $F(X) \subseteq f(X)$, we choose the point $x_1 \in X$ such that $fx_1 \in Fx_0$. If $c = 0$, then

$$d(fx_1, Fx_1) \le H(Fx_0, Fx_1) = 0.$$

Since Fx_1 is compact (hence closed), we obtain $fx_1 \in Fx_1$.

Now we assume $c \neq 0$. By Lemma 2 for each $\varepsilon = \frac{1}{\sqrt{c}} > 1$ there exists a point $y_1 \in Fx_1$ such that

$$d(y_1, Fx_1) \le H(Fx_1, Fx_0) < \varepsilon H(Fx_1, Fx_0).$$

Choose $x_2 \in X$ such that $y_1 = fx_2 \in Fx_1$ and so on. In general, if $x_n \in X$ there exists $x_{n+1} \in X$ such that $y_n = fx_{n+1} \in Fx_n$ and

$$d(y_n, fx_n) < \varepsilon H(Fx_n, Fx_{n-1})$$

for each $n \ge 1$. By the inequality (10) for each $n \in \mathbb{N}$ we have

$$
\begin{aligned}
d(fx_{n+1}, fx_n) &< \varepsilon H(Fx_n, Fx_{n-1}) \leq \frac{c}{\sqrt{c}} \mathcal{N}(F, f) = \sqrt{c}\mathcal{N}(F, f) \\
&< \sqrt{c} \max \{d(fx_n, fx_{n-1}), d(fx_n, Fx_n), d(fx_{n-1}, Fx_{n-1}), \\
&\qquad \frac{1}{2}[d(fx_n, Fx_{n-1}) + d(fx_{n-1}, Fx_n)]\}. \\
&< \sqrt{c} \max \{d(fx_n, fx_{n-1}), d(fx_n, fx_{n+1}), d(fx_{n-1}, fx_n), \\
&\qquad \frac{1}{2} d(fx_{n-1}, fx_{n+1})]\}. \\
&< \sqrt{c} \max \{d(fx_n, fx_{n-1}), d(fx_n, fx_{n+1}), d(fx_{n-1}, fx_n), \\
&\qquad \frac{1}{2}[d(fx_{n-1}, fx_n) + d(fx_n, fx_{n+1})]\}. \\
&= \sqrt{c} \max \{d(fx_{n-1}, fx_n), d(fx_n, fx_{n+1})\}. \\
&= \sqrt{c} d(fx_{n-1}, fx_n)
\end{aligned}
$$

Since $\sqrt{c} < 1$, the sequence $\{fx_n\}$ is a Cauchy sequence on the complete metric space X. Therefore, it converges to a point $z \in X$. Likewise $\{Fx_n\}$ is a Cauchy sequence on the complete metric space $(\mathcal{K}(X), H)$, hence it converges to a set $K \in \mathcal{K}(X)$. As a result

$$
d(z, K) \leq d(z, fx_n) + d(fx_n, K) \leq d(z, fx_n) + H(Fx_{n-1}, K).
$$

Certainly that $d(z, K) = 0$ by $d(z, fx_n) \to 0$ and $H(Fx_{n-1}, K) \to 0$ as $n \to \infty$. This implies $z \in K$ since K is a compact set. Since F and f are compatible, we have

$$
\begin{aligned}
d(fz, Fz) &= \lim_{n \to \infty} d(fz, Fz) \leq \lim_{n \to \infty} [d(fz, ffx_n) + d(ffx_n, Fz)] \\
&\leq \lim_{n \to \infty} [d(fz, ffx_n) + H(fFx_n, Fz)] \\
&\leq \lim_{n \to \infty} [d(fz, ffx_n) + H(fFx_n, Ffx_n) + H(Ffx_n, Fz)] \\
&= d(fz, fz) + H(Fz, Fz) \\
&= 0.
\end{aligned}
$$

So $fz \in Fz$. Conversely, it's clear by Lemma.

Remark 1 Theorem 1.7 is a special occurrence of results obtained by H Kaneko and S Sessa [16]. Certainly the provisioning should be satisfied as in Theorem 3.

This result modify of Theorem 1.5 by substituting compatibility with respect to Hausdorff metric on $\mathcal{K}(X)$ for slight commutativity at once improvement Theorem 1.6 in finding common fixed point for the mapping of the hybrid composite.

Theorem 1.8 Let (X, d) be a complete metric space, $F : X \to \mathcal{K}(X)$ be a set-valued mapping and $f : X \to X$ be a single-valued mapping satisfying the inequality

$$
H(Fx, Fy) \leq \psi \mathcal{N}(F, f) \tag{16}
$$

for all $x, y \in X$, where $\psi : [0, \infty) \to [0, \infty)$ is a nondecreasing, right continuous and $\psi(t) < t$, for all $t > 0$. If this following is satisfied

A. the set-valued mapping F and the single-valued mapping f are continuous,

B. the image $F(X)$ is a subset of the image $f(X)$,

C. the pairs F and f are compatible, and

D. $\exists\, x_0 \in X$ such that $\sup\{H(Fx_n,\ Fx_1) : n = 0, 1\cdots\} < +\infty$,

then F and f have a unique common fixed point on X.

Proof: This proof is the same as Theorem 5 in [14]

Example 1 Let $X = [0, 3]$ with usual metric. Let $F(x) = [0, x^2]$ and $f(x) = 2x^2 - 1$ for each $x \in [0, 3]$. Its clear that the image $F(X) = F([0,\ 3]) = [0, 9] \subset f([0,\ 3]) = [-1, 17] = f(X)$ and the both F and f are continuous on $[0, 3]$. If the sequence $x_n \to 1$, then $Fx_n \to [0, 1] = K$ and $fx_n \to 1 \in K$. We know that $Ffx_n = \left[0,\ (2x_n^2 - 1)^2\right]$ and $fFx_n = \left[-1,\ 2x_n^4 - 1\right]$ so that we obtained

$$H\left(Ffx_n, fFx_n\right) = |4x_n^4 - 6x_n^2 + 2| \to 0$$

since $x_n \to 1$. It is clear $\sup\{H(Fx_n,Fx_1 : n = 0,1,\cdots\} = 9 < +\infty$. Since F and f are continuous, we have

$$\lim_{x_n \to 1} Ffx_n = F(1) = [0,\ 1] = K, \quad and \quad \lim_{x_n \to 1} fx_n = 1 = f(1).$$

This means $1 = f(1) \in F(1) = K$.

Remark 2 Simple examples above prove that the condition of the continuity of the both mappings F and f is important in Theorem 4 other than the other requirements. However, in general the common fixed point theorems for hybrid composite mappings only required one of the mappings F or f is continuous. In our opinion, such case it can be used if the set K is a singleton.

The following main result is a discussion of the existence of a fixed point for the derivative of an interval-valued function.

Theorem 1.9 Suppose that $F : [a, b] \to \mathcal{I}(\mathbb{R})$ is a continuously gh-differentiable on (a, b) such that there exists $f : [a, b] \to \mathbb{R}$ and $fx \in F'_{gh}(x)$ for all $x \in [a, b]$ satisfying the inequality

$$H(Fx,\ Fy) \le \psi \mathcal{N}(F, f) \tag{17}$$

for all $x, y \in [a, b]$, where $\psi : [0, \infty) \to [0, \infty)$ is a non-decreasing, right continuous, and $\psi(t) < t$, for all $t > 0$. If this following is satisfied.

A. the image $F([a,\ b])$ is subsets of the image $f([a,\ b])$,

B. the pairs F and f are compatible, and

C. $\exists\, x_0 \in [a, b]$ such that $\sup\{H(Fx_n,\ Fx_1) : n = 0, 1\cdots\} < +\infty$,

then the gh-derivative F'_{gh} has a unique fixed point.

Proof: From hypothesis (C), suppose $H(Fx_s,\ Fx_t) \le H(Fx_s,\ Fx_1) + H(Fx_t,\ Fx_1) \le M$ so that

$$\sup\{H(Fx_s,\ Fx_t) : s, t = 0,1,2\cdots\} = M < +\infty. \tag{18}$$

Suppose that $N \in \mathbb{N}$ such that for each $\varepsilon > 0$

$$\psi^N L < \varepsilon \tag{19}$$

by Lemma 3.

Let $x_0 \in [a, b]$ be an arbitrary. Since $F([a,\ b]) \subseteq f([a,\ b])$, we choose the point $x_1 \in [a, b]$ such that $y_1 = fx_1 \in Fx_0$. In general, if $x_n \in X$ there exists $x_{n+1} \in X$ such that $y_n = fx_n \in Fx_{n-1}$. By applying inequality (11) to term $H(Fx_m, Fx_n)$ we have for $m,n \geq N$:

$$\begin{aligned} H(Fx_m, Fx_n) &\leq \psi \max \{d(fx_m, fx_n), d(fx_n, Fx_n), d(fx_m, Fx_m), \backslash \\ &\quad \frac{1}{2}[d(fx_n,\ Fx_m) + d(fx_m,\ Fx_n)]\} \\ &\leq \psi \max \{H(Fx_{m-1}, Fx_{n-1}), H(Fx_{n-1}, Fx_n), H(Fx_{m-1}, Fx_m), \\ &\quad \frac{1}{2}[H(Fx_{n-1},\ Fx_m) + H(Fx_{m-1},\ Fx_n)]\} \\ &\leq \psi \max \{H(Fx_{m-1}, Fx_{n-1}), H(Fx_{n-1}, Fx_n), H(Fx_{m-1}, Fx_m), \\ &\quad \frac{1}{2}[H(Fx_{n-1},\ Fx_{m-1}) + H(Fx_{m-1},\ Fx_m)], \\ &\quad \frac{1}{2}[H(Fx_{m-1},\ Fx_{n-1}) + H(Fx_{n-1},\ Fx_n)]\} \\ &= \psi \max \{H(Fx_{m-1},\ Fx_{n-1}),\ H(Fx_{n-1},\ Fx_n),\ H(Fx_{m-1},\ Fx_m)\} \end{aligned} \tag{20}$$

By iterating (14) above as much as N times, we deduce for each $m,n > N$ as follows:

$$\begin{aligned} H(Fx_m, Fx_n) &\leq \psi \max \Big\{H(Fx_r, Fx_s), H(Fx_r, Fx_t), \\ &\quad H(Fx_s, Fx_k) : m - 1 \leq r; t \leq n; n - 1 \leq s; k \leq m\Big\} \\ &\leq \psi^2 \max \{H(Fx_r, Fx_s), H(Fx_r, Fx_t), \\ &\quad H(Fx_s, Fx_k) : m - 2 \\ &\leq r; t \leq n; n - 2 \leq s; k \leq m\} \leq \cdots \\ &\leq \psi^N \max \{H(Fx_r, Fx_s), H(Fx_r, Fx_t), \\ &\quad H(Fx_s, Fx_k) : \qquad 4 \qquad m - N \\ &\leq r; t \leq n; n - N \leq s; k \leq m\} \leq \psi^N M < \varepsilon, \end{aligned} \tag{21}$$

by inequality (13).

Accordingly the sequence $\{Fx_n\}$ is a Cauchy sequence on the complete metric spaces $(\mathcal{I}(\mathbb{R}), H)$ so that converges to an interval $J \in \mathcal{I}(\mathbb{R})$. The sequence of single-valued functions $\{fx_n\}$ is also a Cauchy sequence on $\mathbb{R}$ hence it converges to a point $z \in \mathbb{R}$. We have

$$|z - J| \leq |z - fx_n| + |fx_n - J| \leq |z - fx_n| + H(Fx_{n-1}, J), \tag{22}$$

as $n \to \infty$, $|z - J| = 0$. This means, $z \in J$ since $J \in \mathcal{I}(\mathbb{R})$. By compatibility of F and f, we obtain

$$\lim_{n \to \infty} H(Ffx_n, fFx_n) = 0. \tag{23}$$

By using inequality (11), we have

$$\begin{aligned} H(Ffx_{n+1}, Fx_n) &\leq \psi \max\{d(f^2x_{n+1}, fx_n), d(f^2x_{n+1}, Ffx_{n+1}), d(fx_n, Fx_n), \\ &\quad \frac{1}{2}[d(f^2x_{n+1}, Fx_n) + d(fx_n, Ffx_{n+1})\} \\ &\leq \psi \max\{d(fFx_n, fx_n), d(fFx_n, Ffx_{n+1}), d(fx_n, Fx_n), \\ &\quad \frac{1}{2}[d(fFx_n, Fx_n) + d(fx_n, Ffx_{n+1})\} \\ &\leq \psi \max\{d(fFx_n, Ffx_n) + d(Ffx_n, Fx_n) + d(Fx_n, fx_n), \\ &\quad d(fFx_n, Ffx_n) + d(Ffx_n), fx_n) + d(fx_n, Ffx_{n+1}), \\ &\quad d(fx_n, Fx_n), \frac{1}{2}[d(fFx_n, Fx_n) + d(fx_n, Ffx_{n+1})\} \\ &\leq \psi \max\{d(fFx_n, Ffx_n) + d(Ffx_n, Fx_n) + d(Fx_n, fx_n), \\ &\quad d(fFx_n, Ffx_n) + d(Ffx_n, fx_n) + d(fx_n, Ffx_{n+1})\} \\ &\leq \psi \max\{H(fFx_n, Ffx_n) + H(Ffx_n, Fx_n) + d(Fx_n, fx_n), \\ &\quad H(fFx_n, Ffx_n) + d(Ffx_n, fx_n) + d(fx_n, Ffx_{n+1})\} \\ &\leq \psi \max\{H(fFx_n, Ffx_n) + H(Ffx_n, Fx_n) + d(Fx_n, fx_n), \\ &\quad H(fFx_n, Ffx_n) + H(Ffx_n, Fx_{n-1}) + H(Fx_{n-1}, Ffx_{n+1})\} \end{aligned}$$

since $f^2x_{n+1} \in fFx_n$ and ψ are non-decreasing. Since the pairs F and f are compatible, we obtain

$$\begin{aligned} H(Fz, J) &\leq \psi \max\{0 + H(Fz, J) + d(J, z), 0 + 2H(Fz, J)\} \\ &< \psi \max\{H(Fz, J), 2H(Fz, J)\} \\ &\leq 2\psi H(Fz, J). \end{aligned}$$

Since $\psi(t) < t$ for all $t > 0$, we have $H(Fz, J) = 0$. This means $Fz = J$. Since the pairs F and f are compatible and F is continuously differentiable on $[a, b]$ (hence continuous), we have

$$\lim_{n\to\infty} H(Fz, fJ) = \lim_{n\to\infty} H(Ffx_n, fFx_n) = 0. \tag{24}$$

So $Fz = fJ$. Since $z \in J, f(z) \in f(J)$, consequently

$$f(z) \in F(z) = f(J) = J. \tag{25}$$

Since $f \in F'_{gh}$ and F'_{gh} is continuous, the function f is continuous. Of course, the sequence $\{f^2x_n\}$ converges to the point fz and the sequence of set $\{fFx_n\}$ converges to a set fJ. Since the limit

$$\lim_{n\to\infty} H(Ffx_n, fJ) \leq \lim_{n\to\infty} [H(Ffx_n, fFx_n) + H(fFx_n, fJ)] = 0, \tag{26}$$

the sequence of set $\{Ffx_n\}$ also converges to a set fJ.
Since $f^2x_{n+1} \in fFx_n$ and using inequality (11), we get

$$|f^2x_{n+1} - fx_{n+1})| \leq H(fFx_n, Fx_n) \leq H(fFx_n, Ffx_n) + H(Ffx_n, Fx_n)$$
$$\leq H(fFx_n, Ffx_n) + \psi \max\{|f^2x_n - fx_n|, |f^2x_n - Ffx_n|,$$
$$|fx_n - Fx_n|, \frac{1}{2}[|f^2x_n - Fx_n| + |fx_n, Ffx_n|]\}.$$

For $n \to \infty$, it allows from hypothesis the part B (compatibility) and the Eq. (19) we obtain

$$|fz - z| \leq 0 + \psi \max\left\{|fz - z|, |fz - fJ|, |z - J|, \frac{1}{2}[|fz - J| + |z - Fz|]\right\}$$
$$\leq \psi \max\left\{|fz - z|, 0, 0, \frac{1}{2}[0 + 0]\right\}$$
$$\leq \psi |fz - z|.$$

It implies $z = fz$. Meaning the point z is a fixed point of f. This allows $z = fz \in F(z)$ since the Eq. (19) and hence z is also a fixed point of F'_{gh} by $z = fz \in F'_{gh}(z)$.

Let u is another common fixed point of F and f. By inequality (11), we have that

$$H(Fz, Fu) \leq \psi max\left\{|fz - fu|, |fz - Fz|, |fu - Fu|, \frac{1}{2}[|fz - Fu| + |fu - Fz|]\right\}.$$
$$\leq \psi max\left\{H(Fz, Fu), 0, 0, \frac{1}{2}[H(Fz, Fu) + H(Fu, Fz)]\right\}.$$
$$= \psi max\{H(Fz, Fu)\}.$$
$$= \psi H(Fz, Fu).$$

It implies that $H(Fz, Fu) = 0$. Since $d(z, u) \leq H(Fz, Fu) = 0$, we have $z = u$. Thus the fixed point z is unique. This completes the proof.

Remark 3 To get a common fixed point through the hybrid composite mapping usually contains at least two mappings in its hypothesis. This study shows enough one mapping in its hypothesis. In this case, the mapping given must be differentiable (Theorem 1.9). In addition, the continuity of function is not needed explicitly stated in its hypothesis. Thus this result is more simple than the results reached by past researchers.

Example 2 Let $F(x) = [(x^2 - x), x]$, be an interval-valued function for all $x \in [0, 2]$. It is clear F is gh-differentiable on $(0, 2)$ with derivative

$$F'_{gh}(x) = \begin{cases} [(2x - 1), \ x] & \text{if } 0 \leq x \leq 1, \\ [x, \ (2x - 1)] & \text{if } 1 \leq x \leq 2. \end{cases}$$

In this case, we can take the selector $f(x) = (2x - 1) \in F_{gh}(x)$ for all $x \in [0, 2]$. We obtain the image

$$F([0,\ 2]) = \left[-\frac{1}{4}, 1\right] \cup [1, 2] = \left[-\frac{1}{4}, 2\right] \subset [-1, 2] = f([0,\ 2]).$$

This means that the condition in Theorem 4 part (A) is satisfied.

If the sequence $x_n \to 1$, then $Fx_n \to [0, 1] = K$ and $fx_n \to 1 \in K$. First, we start with the formula $Ffx_n = \left[(2x_n - 1)^2 - (2x_n - 1), (2x_n - 1)\right]$ and $fFx_n = \left[2\left(x_n^2 - x_n\right) - 1, (2x_n - 1)\right]$, we obtain

$$H\left(Ffx_n, fFx_n\right) = |2x_n^2 - 4x_n + 2| \to 0$$

since $x_n \to 1$. Thus F and f are compatible. It is clear that $\sup\{H(Fx_n,\ Fx_1) : n = 0, 1, \cdots\} = 3 < +\infty$. Since F is continuously gh-differentiable on $(0, 2)$, then implies that F and f are continuous on $(0, 2)$ (see Theorem 1.1). Hence we have

$$\lim_{x_n \to 1} Ffx_n = F(1) = [0, 1] = K, \quad \textit{and} \quad \lim_{x_n \to 1} fx_n = 1 = f(1).$$

Certainly $1 = f(1) \in F(1) = K$. Since $f(x) \in F'_{gh}(x)$ for all $x \in [0, 2]$, we obtain $1 = f(1) \in F'_{gh}(1) = 1$. Thus the point $z = 1$ is a unique fixed point of F'_{gh}.

Furthermore, if $f \in F$, then we have the following.

Corollary 1 Suppose that $F : [a, b] \to \mathcal{I}(\mathbb{R})$ is a continuously gh-differentiable on (a, b) such that there exists $f : [a, b] \to \mathbb{R}$ and $fx \in F(x)$ for all $x \in [a, b]$. If the function f and the derivative F'_{gh} satisfies the inequality

$$H\left(F'_{gh}x,\ F'_{gh}y\right) \le \psi\mathcal{N}\left(F'_{gh}, f\right)$$

for all $x,y \in [a, b]$, where $\psi : [0, \infty) \to [0, \infty)$ is a nondecreasing, right continuous, and $\psi(t) < t$, for all $t > 0$ and satisfies the condition.

A. the image $F'([a,\ b])$ is subsets of the image $f([a,\ b])$,

B. the pairs F' and f are compatible, and

C. $\exists\, x_0 \in [a, b]$ such that $\sup\{H(Fx_n,\ Fx_1) : n = 0, 1\cdots\} < +\infty$,

then F'_{gh} has a unique fixed point on X.

Example 3 Let $X = [-2, 2]$ with usual metric. Let $F : [-2, 2] \to \mathcal{I}(\mathbb{R})$ with the formula

$$F(x) = \begin{cases} \left[\left(x + \sin\left(x + \frac{1}{2}\right)\right),\ x\right] & \text{if } -2 \le x \le -\frac{1}{2}, \\ \left[x,\ \left(x + \sin\left(x + \frac{1}{2}\right)\right)\right] & \text{if } -\frac{1}{2} \le x \le -2. \end{cases}$$

It is clear that F is gh-differentiable on $(-2, 2)$ by Theorem 1.2 with derivative

$$F'_{gh}(x) = \begin{cases} \left[1, \left(1+\cos\left(x+\frac{1}{2}\right)\right)\right] & \text{if } -2 \le x \le 1, \\ \left[\left(1+\cos\left(x+\frac{1}{2}\right)\right), 1\right] & \text{if } 1 \le x \le 2. \end{cases}$$

If we choose $f(x) = x \in F(x)$ for all $x \in [-2, 2]$, then we obtain

$$F'_{gh}X = F'_{gh}([-2, 2]) = [1, 2] \cup [0.198, 1] = [0.198, 2] \subset [-2, 2] = f([-2, 2]) = fX.$$

This means the condition in Corollary 1 part (A) is satisfied. If the sequence $x_n \to 1$, then $F'_{gh}(x_n) \to \{1\} = K$ and $fx_n \to 1 \in K$. First, we start with the formula $F'_{gh}fx_n = F'_{gh}(x_n) = \left[\left(1+\cos\left(x_n+\frac{1}{2}\right)\right), 1\right] \cup \left[1, \left(1+\cos\left(x_n+\frac{1}{2}\right)\right)\right]$ and $fF'_{gh}(x_n) = F'_{gh}(x_n) = \left[\left(1+\cos\left(x_n+\frac{1}{2}\right)\right), 1\right] \cup \left[1, \left(1+\cos\left(x_n+\frac{1}{2}\right)\right)\right]$, we obtain

$$H\left(F'_{gh}fx_n, fF'_{gh}x_n\right) = 0.$$

Thus F and f are compatible. Since F is continuously gh-differentiable on $(-2, 2)$, this implies that F and f are continuous on $(-2, 2)$ (see Theorem 1.1). Hence we have

$$\lim_{x_n \to 1} F'_{gh}fx_n = F'_{gh}(1) = \{1\}, \quad \textit{and} \quad \lim_{x_n \to 1} fx_n = 1 = f(1).$$

Consequently, $1 = f(1) \in F'_{gh}(1) = \{1\}$.

5. Conclusions

The existence of a fixed point for the derivative of set-valued mappings can be obtained by using the method of the compatibility of the hybrid composite mappings in the sense of the Pompei-Hausdorff metric.

Conflict of interest

The authors declare no conflict of interest.

Author details

Mohamad Muslikh[1*†] and Adem Kilicman[2†]

1 Universitas Brawijaya, Malang, Indonesia

2 Universiti Putra malaysia, Serdang, Malaysia

*Address all correspondence to: mslk@ub.ac.id

† These authors contributed equally.

References

[1] Baxter G. Mint: On fixed points of the composite of commuting functions. 1963. Available from: http://www.ams.org/journal-terms-of-use. [Accessed: 2018-12-31]

[2] Boyce WM. Mint: Commuting functions with no common fixed point. 1967. Available from: http://www.ams.org/journal-terms-of-use. [Accessed: 2017-10-3]

[3] Husein SA, Sehgal VM. Mint: On common fixed point for a family of mappings. Bulletin of the Australian Mathematical Society. 1975;**13**:261-267. [Accessed: 2018-12-31]

[4] Singh SL, Meade BA. Mint: On common fixed point theorems. Bulletin of the Australian Mathematical Society. 1977;**16**:49-53. [Accessed: 2018-12-31]

[5] Jungck G. Mint: Commuting mappings and fixed points. The American Mathematical Monthly. 1976; **83**(4):261-263. Available from: http://www.jstor.org/stable/2318216. [Accessed: 2017-10-31]

[6] Mehta JG, Joshi ML. Mint: On common fixed point theorem in complete metric spaces. General Mathematics Notes. 2012;**2**(1):55-63

[7] Fisher B. Mint: Result on common fixed points on complete metric spaces. Glasgow Mathematical Journal. 1980;**21**: 65-67

[8] Khan MS. Mint: Common fixed point theorems for multivalued mappings. Pacific Journal of Mathematics. 1981; **95**(2):337-347

[9] Abdou AAN. Mint: Common fixed point result for multi-valued mappings with some examples. Journal of Nonlinear Sciences and Applications. 2016;9:787-798. Available from: www.tjnsa.com [Accessed: 2017-07-31]

[10] Fisher B. Mint: Common fixed point theorems for mappings and set-valued mappings. Rostocker Mathematisches Kolloquium. 1981;**18**:69-77

[11] Singh SL, Kamal R, Sen MDL, Chugh R. Mint: New a type of coincidence and common fixed point theorem with applications. Abstract and Apllied Analysis. 2014;**2014**:1-11. [Accessed: 2017-08-31]

[12] Itoh S, Takahashi W. Mint: Single-valued mappings, set-valued mappings and fixed point theorrems. Journal of Mathematical Analysis and Applications. 1977;**59**:514-521

[13] Fisher B. Mint: Fixed point of mappings and set-valued mappings. The Journal of the University of Kuwait, Science. 1982;**9**:175-180

[14] Imdad M, Khan MS, Sessa S. On some weak conditions of commutativity in common fixed point theorems. International Journal of Mathematics and Mathematical Sciences. 1988;**11**(2): 289-296

[15] Sessa S, Khan MS, Imdad M. Mint: Common fixed point theorem with a weak commutativity condition. Glasnik Mathematicki. 1986;**21**(41):225-235

[16] Kaneko H, Sessa S. Mint: Fixed point theorems for multivalued and single mappings. International Journal of Mathematics and Mathematical Sciences. 1989;**12**(2):257-262

[17] Jungck G, Rhoades BE. Mint: Some fixed point theorems for compatible maps. International Journal of

Mathematics and Mathematical Sciences. 1993;**16**(3):417-428

[18] Elekes M, Keleti T, Prokaj V. Mint: The fixed point of the composition of derivatives. Real Analysis Exchange. 2001-2002;**27**(1):131-140

[19] Muslikh M, Kilicman A. On common fixed point of a function and its derivative. Advances Fixed Point Theory. 2017;7(3):359-371

[20] Pompeiu D. Sur a Continuité des fonctions devariables complexes, [Theses]. Paris: Gauthier-Villars; 1905

[21] Hausdorff F. Grundzughe der Mengenlehre. Leipzig: Viet; 1914

[22] Moore RE, Kearfott RB, Cloud MJ. Introduction to Interval Analysis. Philadelphia, USA: The society for Industrial and applied Mathematics; 2009

[23] Hukuhara M. Intégration des applications mesurables dont la valeur est uncompact convex. Fako de l'Funkcialaj Ekvacioj Japana Matematika Societo. 1967;**10**:205-229

[24] Markov S. Calculus for Interval Function of a real variables. Computing. 1969;**22**:325-337

[25] Stefanini L. A generalization of Hukuhara difference and division for interval and fuzzy arithmetic. Fuzzy Sets and System. 2010;**161**(11):1564-1584

[26] Stefanini L, Bede B. Generalized Hukuhara differentiability of interval valued functions and interval differential equations. Nonlinear Analysis. 2009;**71**: 1311-1328

[27] Chalcao-Chano Y, Machui-Huaman GG, Silva GN, Jimenez-Gamero MD. Algebra of generalized Hukuhara differentiable interval-valued functions: Review and new properties. Fuzzy Sets and System. 2019;**375**:53-69

[28] Fisher B. Common fixed point theorems for mappings and set-valued mappings. The Journal of the University of Kuwait, Science. 1984;**11**:15-21

[29] Nadler SB Jr. Multivalued contraction mappings. Pacific Journal of Mathematics. 1969;**30**(2):475-488

Chapter 6

Stability Estimates for Fractional Hardy-Schrödinger Operators

Konstantinos Tzirakis

Abstract

In this chapter, we derive optimal Hardy-Sobolev type improvements of fractional Hardy inequalities, formally written as $\mathcal{L}_s u \geq \frac{w(x)}{|x|^\theta} u^{2_*-1}$, for the fractional Schrödinger operator $\mathcal{L}_s u = (-\Delta)^s u - k_{n,s} \frac{u}{|x|^{2s}}$ associated with s-th powers of the Laplacian for $s \in (0, 1)$, on bounded domains in $\mathbb{R}^n$. Here, $k_{n,s}$ denotes the optimal constant in the fractional Hardy inequality, and $2_* = \frac{2(n-\theta)}{n-2s}$, for $0 \leq \theta \leq 2s < n$. The optimality refers to the singularity of the logarithmic correction w that has to be involved so that an improvement of this type is possible. It is interesting to note that Hardy inequalities related to two distinct fractional Laplacians on bounded domains admit the same optimal remainder terms of Hardy-Sobolev type. For deriving our results, we also discuss refined trace Hardy inequalities in the upper half space which are rather of independent interest.

Keywords: fractional Laplacian, hardy-Sobolev inequalities, Schrödinger operator

1. Introduction

Fractional Laplacian operators have attracted considerable attention in various areas of pure and applied mathematics, see for instance [1] and the review articles [2–4]. Such non-local operators appear naturally in several branches of the applied sciences to model phenomena where long-range interactions take place, in fluid dynamics, quantum mechanics, biological populations, materials science, finance, image processing, and game theory, to name a few, for example, [5–16]. They have a prominent interest from a mathematical point of view, arising in analysis and partial differential equations (pdes), geometry, probability, and financial mathematics, see for instance [17–22].

For $0 < s < 1$, the fractional Laplacian $(-\Delta)^s$ of a function f in the Schwartz space of rapidly decaying C^∞ functions on $\mathbb{R}^n$, is defined as a pseudodifferential operator (e.g., [1, 23, 24])

$$(-\Delta)^s f = \mathcal{F}^{-1}\Big(|\xi|^{2s}(\mathcal{F}f)\Big), \qquad \forall \xi \in \mathbb{R}^n, \tag{1}$$

where, $\mathcal{F}f$ denotes the Fourier transform of f defined by

IntechOpen

$$\mathcal{F}f(\xi) = \frac{1}{(2\pi)^{n/2}} \int_{\mathbb{R}^n} e^{-i\xi\cdot x} f(x)\,dx.$$

It can be shown that the operator $(-\Delta)^s$ can be equivalently defined as the singular integral operator (see for instance [1], Proposition 3.3])

$$\begin{aligned}(-\Delta)^s f(x) &= c(n,s)\text{P.V.}\int_{\mathbb{R}^n} \frac{f(x)-f(y)}{|x-y|^{n+2s}}\,dy \\ &:= c(n,s)\lim_{\varepsilon\to 0^+}\int_{\{|x-y|>\varepsilon\}} \frac{f(x)-f(y)}{|x-y|^{n+2s}}\,dy, \qquad \forall x \in \mathbb{R}^n,\end{aligned} \tag{2}$$

where

$$c(n,s) = \frac{s4^s}{\pi^{n/2}} \frac{\Gamma\left(\frac{n+2s}{2}\right)}{\Gamma(1-s)} \tag{3}$$

and Γ stands for the usual Gamma function defined by $\Gamma(s) = \int_0^\infty t^{s-1}e^{-t}dt$. Notice that, if $s < 1/2$, then the integrand exhibits an integrable singularity, thus the principal value (P.V.) may be dropped. Moreover, by a change of variable, we can avoid the principal value and transform the singular integral in (2) as

$$(-\Delta)^s f(x) = \frac{1}{2} c(n,s) \int_{\mathbb{R}^n} \frac{2f(x) - f(x+y) - f(x-y)}{|y|^{n+2s}}\,dy.$$

We caution the reader to take into account the conventional value imposed for the constant $c(n,s)$ when comparing different definitions for fractional Laplacian. Here, we fix the value (3) so that the singular integral representation (2) accords with the characterization (1) as a Fourier multiplier operator, and notice that $\lim_{s\to 1^-}(-\Delta)^s f = -\Delta f$ and $\lim_{s\to 0^+}(-\Delta)^s f = f$. Note that the definition (1) allows for a wider range of the fractional Laplace's exponents s, while the expression (2) is defined for $s < 1$. We point out that the characterization via Fourier transform is reduced to the standard Laplacian as $s \to 1$, which, however cannot be defined by the pointwise expression (2). Let us also remark that from the definition in the Schwartz space it is possible to extend $(-\Delta)^s$ by duality in a large class of tempered distributions; see, for example [25]. For a further discussion on the fractional Laplacian and the associated fractional Sobolev spaces we refer the readers to ([1], §§2–3]).

In the literature, other characterizations for $(-\Delta)^s$ are also used, that turn out to be equivalent to the definitions (1), (2). A further discussion on the different definitions of the fractional Laplacian on $\mathbb{R}^n$ and a proof of their equivalence can be found in [26]. Each of these equivalent characterizations allows for different approaches for the related problems, and in our context, we exploit a characterization realizing the nonlocal operator via an appropriate extended local problem (see Section 3), where local pdes techniques can be applied.

Regarding the corresponding quadratic form for $(-\Delta)^s$,

$$((-\Delta)^s f, f) := \int_{\mathbb{R}^n} f\,(-\Delta)^s f\;dx = \int_{\mathbb{R}^n} |\xi|^{2s}\,(\mathcal{F}f)^2(\xi)\;d\xi$$

we have (see Aronszajn-Smith [27], page 402)

$$\int_{\mathbb{R}^n} |\xi|^{2s} (\mathcal{F}f)^2(\xi)\ d\xi = \frac{c(n,s)}{2} \int_{\mathbb{R}^n} \int_{\mathbb{R}^n} \frac{|f(x) - f(y)|^2}{|x-y|^{n+2s}}\ dx\ dy. \tag{4}$$

We consider the homogeneous fractional Sobolev space $\dot{H}^s(\mathbb{R}^n)$, defined as the completion of $C_0^\infty(\mathbb{R}^n)$ with respect to

$$\|f\|_{\dot{H}^s(\mathbb{R}^n)} := \int_{\mathbb{R}^n} \int_{\mathbb{R}^n} \frac{|f(x) - f(y)|^2}{|x-y|^{n+2s}}\ dx\ dy. \tag{5}$$

The sharp fractional Sobolev inequality, associated to $(-\Delta)^s$, states that

$$S_{n,s} \left(\int_{\mathbb{R}^n} |f|^{2_s^*}(x)\ dx \right)^{2/2_s^*} \leq \int_{\mathbb{R}^n} \int_{\mathbb{R}^n} \frac{|f(x) - f\ \ (y)|^2}{|x-y|^{n+2s}}\ dx\ dy, \quad \forall f \in \dot{H}^s(\mathbb{R}^n), \tag{6}$$

where $2_s^* = \frac{2n}{n-2s}$, and the best constant

$$S_{n,s} = \frac{2^{2s}\pi^s \Gamma\left(\frac{n+2s}{2}\right)}{\Gamma\left(\frac{n-2s}{2}\right)} \left(\frac{\Gamma\left(\frac{n}{2}\right)}{\Gamma(n)} \right)^{2s/n}$$

is achieved in $\dot{H}^s(\mathbb{R}^n)$, exactly by the multiples, dilates, and translates of the function $\left(1 + |x|^2\right)^{(2s-n)/2}$; see [28, 29]. Sobolev inequality (6) yields the continuous embedding $\dot{H}^s(\mathbb{R}^n) \hookrightarrow L^{2_s^*}(\mathbb{R}^n)$, which is sharp within the framework of Lebesgue spaces, in the sense that the embedding fails for any other Lebesgue subspace. In terms of Lorentz spaces, this embedding reads as $\dot{H}^1(\mathbb{R}^n) \hookrightarrow L^{2_s^*, 2_s^*}(\mathbb{R}^n)$, which admits an extension within the whole Lorentz space scale $L^{2_s^*, p}(\mathbb{R}^n)$, $p \geq 2$. As a matter of fact, the embeddings for $p > 2$, follow from the continuous inclusions $L^{2_s^*, 2}(\mathbb{R}^n) \hookrightarrow L^{2_s^*, p}(\mathbb{R}^n)$, and the continuous embedding

$$\dot{H}^s(\mathbb{R}^n) \hookrightarrow L^{2_s^*, 2}(\mathbb{R}^n), \tag{7}$$

which, in turn, follows from the fractional Hardy inequality

$$k_{n,s} \int_{\mathbb{R}^n} \frac{|f(x)|^2}{|x|^{2s}}\ dx \leq \int_{\mathbb{R}^n} \int_{\mathbb{R}^n} \frac{|f(x) - f\ \ (y)|^2}{|x-y|^{n+2s}}\ dx\ dy. \tag{8}$$

Indeed, one can derive (7) from (8), by the fact that under radially decreasing rearrangement the $\dot{H}^s(\mathbb{R}^n)$ norm does not increase [30] and the left hand side of (8) does not decrease, while the Lorentz quasinorm $\|\cdot\|_{L^{2_s^*, 2}}$ is invariant and proportional to the left hand side of (8).

In this sense, Hardy's inequality (8) is stronger than Sobolev's inequality (6). The value

$$k_{n,s} = \frac{2\pi^{n/2}\Gamma(1-s)\Gamma^2\left(\frac{n+2s}{4}\right)}{s\Gamma^2\left(\frac{n-2s}{4}\right)\Gamma\left(\frac{n+2s}{2}\right)}$$

is the best possible constant in (8). It is well known that the best constant $k_{n,s}$ in (8) is not attained in $\dot{H}^s(\mathbb{R}^n)$, yet no L^p improvement is possible in $\dot{H}^s(\mathbb{R}^n)$, as demonstrated by testing with suitable perturbations of the solution $|x|^{\frac{2s-n}{2}}$, of the corresponding Euler–Lagrange equation.

An application of Hölder's inequality together with (6) and (8), yield the following Hardy-Sobolev inequality:

$$\Lambda_{n,\theta,s}\int_{\mathbb{R}^n}\frac{|f|^{2_*(\theta)}}{|x|^{\theta}}\,dx \le \int_{\mathbb{R}^n}\int_{\mathbb{R}^n}\frac{|f(x)-f\ (y)|^2}{|x-y|^{n+2s}}\,dx\ dy,\quad f\in C_0^{\infty}(\mathbb{R}^n), \tag{9}$$

where $2_*(\theta)=\frac{2(n-\theta)}{n-2s}$, $0\le\theta<2s$. The best constant in (9), contrary to the border-line case (8) i.e. $\theta=2s$, is achieved in $\dot{H}^s(\mathbb{R}^n)$; cf. [31].

In view of (3)–(4), inequality (8) is equivalent to

$$h_{n,s}\int_{\mathbb{R}^n}\frac{f^2(x)}{|x|^{2s}}\,dx \le \int_{\mathbb{R}^n}|\xi|^{2s}(\mathcal{F}f)^2(\xi)\ d\xi,\quad \forall f\in\dot{H}^s(\mathbb{R}^n), \tag{10}$$

with the sharp constant

$$h_{n,s}=4^s\Gamma^2\left(\frac{n+2s}{4}\right)/\Gamma^2\left(\frac{n-2s}{4}\right). \tag{11}$$

The dual form of (10), formulated in terms of Riesz integral operator, is a special case of Stein-Weiss inequalities [32], and the best constant $h_{n,s}$ is identified by Herbst [33]; see also Beckner [34], Yafaev [35].

By Hardy-Littlewood and Pólya-Szegö type rearrangement inequalities, it suffices to prove (10) for radial decreasing f; see Almgren and Lieb [30] where it is shown that (4) does not increase if f is replaced by its equimeasurable symmetric decreasing rearrangement. Then, we will show that the inequality is equivalent to a convolution inequality on the multiplicative group $\mathbb{R}_+$ equipped with the Haar measure $\frac{1}{r}dr$.

In particular, (10) is equivalent to the following doubly weighted Hardy-Littlewood-Sobolev inequality of Stein-Weiss [32].

$$\int_{\mathbb{R}^n}\int_{\mathbb{R}^n}\frac{f(x)}{|x|^s}\,\frac{1}{|x-y|^{n-2s}}\,\frac{f(y)}{|y|^s}\,dxdy\ \le\ C_{n,s}\int_{\mathbb{R}^n}|f(x)|^2\,dx, \tag{12}$$

with sharp constant

$$C_{n,s}=\frac{\pi^{n/2}\Gamma^2\left(\frac{n-2s}{4}\right)\Gamma(s)}{\Gamma^2\left(\frac{n+2s}{4}\right)\Gamma\left(\frac{n-2s}{2}\right)}.$$

Since we can assume that f is radial, we set $f(x)=\mathrm{f}(r)$, and $x=rx'$, $y=\rho y'$ where $|x'|=|y'|=1$. Regarding the convolution integral of the left side in (12), we employ polar coordinates to get

$$
\begin{aligned}
&\int_{\mathbb{R}^n}\int_{\mathbb{R}^n} f(x)\ \frac{1}{|x|^s}\ \frac{1}{|x-y|^{n-2s}}\ \frac{1}{|y|^s}\ f(y)\ dxdy\ = \\
&\int_0^\infty\int_0^\infty \int_{|x'|=1}\int_{|y'|=1} \mathrm{f}(r)\ \frac{r^{n-1}}{r^s}\ \frac{1}{|rx'-\rho y'|^{n-2s}}\ \frac{\rho^{n-1}}{\rho^s}\ \mathrm{f}(\rho)\ d\sigma(x')d\sigma(y')drd\rho\ = \\
&\int_0^\infty\int_0^\infty \int_{|x'|=1} \left[\mathrm{f}(r)r^{n/2}\right]\ \frac{1}{r^{\frac{2s-n}{2}}}\ K(r,\rho)\ \frac{1}{\rho^{\frac{2s-n}{2}}}\ \left[\mathrm{f}(\rho)\rho^{n/2}\right]\ d\sigma(x')\ \frac{dr}{r}\frac{d\rho}{\rho}
\end{aligned}
\tag{13}
$$

where $d\sigma$ denotes $(n-1)$-dimensional Lebesgue integration over the unit sphere $\mathbb{S}^{n-1} = \{x' \in \mathbb{R}^n :\ |x'| = 1\}$, and we set

$$
K(r,\rho) := \int_{|y'|=1} \frac{1}{|rx'-\rho y'|^{n-2s}}\ d\sigma(y'). \tag{14}
$$

Notice that $K(r,\rho)$ in (14) is independent of $x' \in \mathbb{S}^{n-1}$. To show this independence, we may assume $r = 1, \rho = \tau$, or more generally, to use the variable $\tau = \rho/r$ and then it suffices to show that

$$
K(\tau) := \int_{|y'|=1} \frac{1}{|x'-\tau y'|^{n-2s}} d\sigma(y')
$$

is independent of $x' \in \mathbb{S}^{n-1}$. Indeed, take an arbitrary $z' \in \mathbb{S}^{n-1}$. Then there exists a rotation R such that $z' = Rx'$ and we denote by R^T its transpose. Performing the change of variables $w' = R^T y'$, we get

$$
\int_{|y'|=1} \frac{1}{|z'-\tau y'|^{n-2s}} d\sigma(y') = \int_{|w'|=1} \frac{1}{|x'-\tau w'|^{n-2s}} d\sigma(w') = K(\tau),
$$

since $|detR| = 1$ and $|Rv_1 - Rv_2| = |v_1 - v_2|$, for every $v_1, v_2 \in \mathbb{R}^n$. Since $K(r,\rho)$ is independent of $x' \in \mathbb{S}^{n-1}$ we have

$$
\int_{\mathbb{S}^{n-1}} K(r,\rho) d\sigma(x')\ =\ K(r,\rho)\int_{\mathbb{S}^{n-1}} 1 d\sigma(x')\ =\ K(r,\rho)\,\frac{2\pi^{n/2}}{\Gamma\left(\frac{n}{2}\right)}. \tag{15}
$$

Moreover, in (14), we can choose x' to be the first direction unit vector in $\mathbb{R}^n$, that is $\hat{e}_1 = (x_1, x_2, \cdots, x_n)$ with $x_1 = 1, x_2 = x_3 = \cdots = x_n = 0$, hence

$$
K(r,\rho) = \int_{|y'|=1} \frac{1}{\left(r^2 - 2r\rho y_1 + \rho^2\right)^{\frac{n-2s}{2}}}\ d\sigma(y')
$$

thus

$$
\frac{1}{r^{\frac{2s-n}{2}}}\ K(r,\rho)\ \frac{1}{\rho^{\frac{2s-n}{2}}} = \int_{|y'|=1} \frac{1}{\left(\frac{r}{\rho} - 2y_1 + \frac{\rho}{r}\right)^{\frac{n-2s}{2}}}\ d\sigma(y')
$$

and substituting (15) into (13), we get

$$\int_{\mathbb{R}^n}\int_{\mathbb{R}^n} f(x)\ \frac{1}{|x|^s}\ \frac{1}{|x-y|^{n-2s}}\ \frac{1}{|y|^s}\ f(y)\ dx\,dy = \frac{2\pi^{n/2}}{\Gamma\left(\frac{n}{2}\right)}\int_0^\infty\int_0^\infty h(r)\ \psi\left(\frac{r}{\rho}\right)\ h(\rho)\ \frac{dr}{r}\frac{d\rho}{\rho} \tag{16}$$

where

$$h(r) := \mathrm{f}(r)r^{n/2} \quad \text{and} \quad \psi(\tau) = \psi\left(\frac{1}{\tau}\right) = \int_{|y'|=1} \frac{1}{\left(\tau - 2y_1 + \frac{1}{\tau}\right)^{\frac{n-2s}{2}}}\ d\sigma(y').$$

As for the right side of the fractional integral inequality (12), we use again polar coordinates to get

$$\int_{\mathbb{R}^n} |f(x)|^2\ dx \ = \ \frac{2\pi^{\frac{n}{2}}}{\Gamma\left(\frac{n}{2}\right)}\int_0^\infty |h(r)|^2\frac{dr}{r}. \tag{17}$$

Finally, substituting (16), (17) in (12), we conclude that the fractional Hardy inequality (10) is written equivalently as the convolution inequality

$$\int_0^\infty\int_0^\infty h(r)\ \psi\left(\frac{r}{\rho}\right)\ h(\rho)\ \frac{dr}{r}\frac{d\rho}{\rho} \le C_{n,s}\int_0^\infty |h(r)|^2\frac{dr}{r}. \tag{18}$$

Inequality (18) is a convolution inequality on the multiplicative group $\mathbb{R}_+$ equipped with the Haar measure $\frac{1}{r}dr$, and using the sharp Young's inequality for convolution on certain noncompact Lie groups, we recover the sharpness of the constant and the non-existence of extremals for the fractional Hardy inequality (10).

2. Fractional hardy-Sobolev inequalities on bounded domains

In the sequel, we will discuss Hardy type inequalities for fractional powers of Laplacian associated with bounded domains, and, more precisely, defined for functions satisfying homogeneous Dirichlet boundary or exterior conditions. So hereafter let us fix a bounded domain $\Omega \subset \mathbb{R}^n$, with $n > 2s$.

In opposition to the case of the whole of $\mathbb{R}^n$, distinct definitions of such non-local operators have been introduced as mathematical models in various applications. In particular, we consider two of the most commonly used operators of this type, which are the so-called spectral Laplacian (see e.g. [36–38] and references therein) and the Dirichlet (also referred to as *restricted* or *regional* or integral, see e.g. [39, 40], and references therein). Both operators are deeply associated with the theory of stochastic processes. They can be characterized as generators of a $(2s)$-stable Lévy process with jumps resulting from two consecutive modifications of Wiener process, the subordination and the stopping (killing the process when leaves the domain), which reflect the homogeneous Dirichlet-type boundary (or exterior) conditions. Depending on which of these modifications is first applied, we take two different stochastic processes and their corresponding infinitesimal generators.

The Dirichlet fractional Laplacian Next, we will discuss improved versions of fractional Hardy inequalities, involving sharp Sobolev-Hardy type correction terms.

We begin with the Dirichlet fractional Laplacian which we again denote by $(-\Delta)^s$. We merely extend any function $f \in C_0^\infty(\Omega)$ in the entire $\mathbb{R}^n$ by defining $f(x) = 0$, for any $x \notin \Omega$, and then we define $(-\Delta)^s f$ as the standard fractional Laplacian on the whole space, acting on the extension of f to $\mathbb{R}^n$. More precisely, we define

$$(-\Delta)^s f = \mathcal{F}^{-1}\left(|\xi|^{2s}(\mathcal{F}f)\right), \qquad \forall \xi \in \mathbb{R}^n.$$

The Dirichlet fractional Laplacian can be equivalently characterized as the singular integral operator (2) for the $c(n,s)$ given in (3).

Passing from $\mathbb{R}^n$ to a bounded domain Ω, containing the origin, inequality (8) is still valid with the same best possible constant

$$k_{n,s}\int_\Omega \frac{f^2(x)}{|x|^{2s}}\,dx \le \int_{\mathbb{R}^n}\int_{\mathbb{R}^n} \frac{|f(x)-f(y)|^2}{|x-y|^{n+2s}}\,dx\,dy, \qquad \forall f \in H_0^s(\Omega), \tag{19}$$

where $H_0^s(\Omega)$ is the homogeneous fractional Sobolev space, defined as the completion of the functions in $C_0^\infty(\Omega)$, extended by zero outside Ω, with respect to the norm (5). Clearly the constant $k_{n,s}$ can not be achieved in $H_0^s(\Omega)$, and various improved versions of (19) have been established by many authors, which amount to adding L^p norms of u or its fractional gradients in the left hand side.

In particular, Frank, Lieb and Seiringer have shown among others in [40], that for any $1 \le q < 2_s^* := 2n/(n-2s)$ and any bounded domain $\Omega \subset \mathbb{R}^n$ there exists a positive constant $c = c(n,s,q,\ |\Omega|)$ such that

$$k_{s,n}\int_\Omega \frac{f^2(x)}{|x|^{2s}}\,dx + c\left(\int_\Omega |f(x)|^q\,dx\right)^{2/q} \le \int_{\mathbb{R}^n}\int_{\mathbb{R}^n} \frac{|f(x)-f(y)|^2}{|x-y|^{n+2s}}\,dx\,dy, \qquad f \in C_0^\infty(\Omega). \tag{20}$$

Using the Dirichlet to Neumann mapping for the representation of the fractional Laplacian [39] (see Section 3 for details), a partial extension of (20) has been obtained in [41], replacing the remainder term with the $p-$norm of a fractional gradient, $p < 2$.

An improvement involving a 2-norm of a fractional gradient, has been obtained in [42], using the following representation of the remainder term ([40], Proposition 4.1),

$$\begin{aligned} &k_{n,s}\int_{\mathbb{R}^n} \frac{f^2(x)}{|x|^{2s}}\,dx - \int_{\mathbb{R}^n}\int_{\mathbb{R}^n} \frac{|f(x)-f(y)|^2}{|x-y|^{n+2s}}\,dx\,dy \\ &= c(n,s)\int_{\mathbb{R}^n}\int_{\mathbb{R}^n} \frac{|v(x)-v(y)|^2}{|x-y|^{n+2s}}\frac{1}{|x|^{\frac{n-2s}{2}}}\frac{1}{|y|^{\frac{n-2s}{2}}}\,dx\,dy \end{aligned} \tag{21}$$

with the ground state substitution

$$v(x) = f(x)|x|^{\frac{n-2s}{2}}. \tag{22}$$

We point out that the exponent q in (20) is strictly smaller than the critical fractional Sobolev exponent 2_s^* and the inequality fails for $q = 2_s^*$. In [43] we have shown that introducing a logarithmic relaxation we can have a critical Sobolev improvement of (19). More precisely, it has been shown the existence of a positive constant C, depending only on n and s, such that for $f \in H_0^s(\Omega)$,

$$k_{n,s}\int_\Omega \frac{|f(x)|^2}{|x|^{2s}}\,dx + C\left(\int_\Omega X^{\frac{2(n-s)}{n-2s}}\left(\frac{|x|}{D}\right)|f(x)|^{\frac{2n}{n-2s}}\,dx\right)^{\frac{n-2s}{n}} \le \int_{\mathbb{R}^n}\int_{\mathbb{R}^n}\frac{|f(x)-f(y)|^2}{|x-y|^{n+2s}}\,dx\,dy, \tag{23}$$

where $D = \sup_{x\in\Omega}|x|$ and

$$X(r) = (1-\ln r)^{-1}, \quad 0 < r \le 1.$$

Moreover, the weight $X^{\frac{2(n-s)}{n-2s}}$ cannot be replaced by a smaller power of X. We emphasize that inequality (23) involves the critical exponent but contrary to the subcritical case, that is (20), it has a logarithmic correction. However inequality (23) is sharp in the sense that inequality fails for smaller powers of the logarithmic correction X. This result may be seen as the fractional version of (see [44, 45])

$$\frac{(n-2)^2}{4}\int_\Omega \frac{|f(x)|^2}{|x|^2}\,dx + c_n\left(\int_\Omega |f(x)|^{\frac{2n}{n-2}}X^{\frac{2(n-1)}{n-2}}(|x|/D)\,dx\right)^{\frac{n-2}{n}} \le \int_\Omega |\nabla f|^2\,dx, \tag{24}$$

in the sense that (23) reduces to (24) when $s \to 1^-$.

Moreover, in [43] we have shown, for some constant $C > 0$,

$$k_{n,s}\int_\Omega \frac{|f(x)|^2}{|x|^{2s}}\,dx + C\int_\Omega X^2\left(\frac{|x|}{D}\right)|f(x)|^2 dx \le \int_{\mathbb{R}^n}\int_{\mathbb{R}^n}\frac{|f(x)-f(y)|^2}{|x-y|^{n+2s}}\,dx\,dy, \tag{25}$$

where the weight X^2 cannot be replaced by a smaller power of X.

Let us notice that contrary to the Hardy-Sobolev inequalities obtained in [46], where the Hardy potential entails the distance to the boundary, the Hardy-Sobolev inequalities involving the distance from the origin, miss the critical-Sobolev exponent by a logarithmic correction which cannot be removed. Let us also emphasize that our results cover the full range $s \in (0,1)$, in contrast to the case involving the distance from the boundary, where Hardy inequalities associated with the spectral and Dirichlet fractional Laplacians fail within the range $0 < s < 1/2$.

In view of (23) and (25), we can apply Hölder inequality to get the following Hardy-Sobolev improvement of (19).

Theorem 1. *Let $s \in (0,1)$, $0 \le \theta \le 2s$, Ω be a bounded domain in $\mathbb{R}^n$ with $n > 2s$. Then there exists a positive constant $C = C(n,s,\theta)$ such that*

$$h_{n,s}\int_\Omega \frac{|f(x)|^2}{|x|^{2s}}\,dx + C\left(\int_\Omega \frac{X^{p(\theta)}}{|x|^\theta}|f|^{2_*(\theta)}dx\right)^{\frac{2}{2_*(\theta)}} \le ((-\Delta)^s f, f),$$

for any $f \in C_0^\infty(\Omega)$, or equivalently,

$$k_{n,s}\int_\Omega \frac{|f(x)|^2}{|x|^{2s}}\,dx + C\left(\int_\Omega \frac{X^{p(\theta)}}{|x|^\theta}|f|^{2_*(\theta)}dx\right)^{\frac{2}{2_*(\theta)}} \le \int_{\mathbb{R}^n}\int_{\mathbb{R}^n}\frac{|f(x)-f(y)|^2}{|x-y|^{n+2s}}\,dx\,dy, \tag{26}$$

where $2_*(\theta) = \frac{2(n-\theta)}{n-2s}$, $p(\theta) = \frac{2n-\theta-2s}{n-2s}$ and $X = X(|x|/D)$ with $D = \sup_{x\in\Omega}|x|$. The logarithmic weight cannot be replaced by a smaller power of X.

The optimality of the exponent $p := p(\theta) = \frac{2(n-s)-\theta}{n-2s}$ of the logarithmic weight, for the range $\theta \in [0, 2s)$ can be deduced by the optimality of the exponent of the weight X^2, for the case $\theta = 2s$, jointly with Hölder inequality; cf. ([43], Remark), [47].

In view of (21), under the substitution (22) inequality (26) yields sharp limiting cases of certain fractional Caffarelli-Kohn-Nirenberg inequalities established in [48, 49].

The spectral fractional Laplacian We proceed with another reasonable approach in defining a nonlocal operator related to fractional powers of the Laplacian on the bounded domain Ω. We consider an orthonormal basis of $L^2(\Omega)$, consisting of eigenfunctions of $-\Delta$ with homogeneous Dirichlet boundary conditions, say $\phi_1, \dots, \phi_k, \dots$, with corresponding eigenvalues

$$0 < \lambda_1 < \lambda_2 \le \lambda_3 \le \cdots \qquad \text{with} \qquad \lambda_k \to \infty.$$

More precisely,

$$\begin{cases} -\Delta\phi_k = \lambda_k\phi_k, & \text{in } \Omega, \\ \phi_k = 0, & \text{on } \partial\Omega. \end{cases}$$

Then we have.

$$f = \sum_{k=1}^{\infty} c_k\phi_k \qquad \text{where} \qquad c_k = \int_\Omega f\phi_k \, dx.$$

For any $0 < s < 1$, the spectral fractional Laplacian, denoted hereafter by A_s, is defined, similarly to the spectral decomposition of the standard Laplacian, by

$$A_s f = \sum_{k=1}^{\infty} \lambda_k^s c_k \phi_k, \quad \forall f \in C_0^\infty(\Omega).$$

Notice that the operator A_s can be extended by approximation for functions in the Hilbert space

$$H = \left\{ f = \sum_{k=1}^{\infty} c_k\phi_k \in L^2(\Omega) : \|f\|_H = \left(\sum_{k=1}^{\infty} \lambda_k^s c_k^2\right)^{1/2} < \infty \right\}.$$

The quadratic form corresponding to A_s is given by

$$(A_s f, f) := \int_\Omega f A_s f \, dx = \sum_{k=1}^{\infty} \lambda_k^s c_k^2.$$

Let us point out that, contrary to the case of the whole space $\mathbb{R}^n$, the fractional operators A_s and $(-\Delta)^s$, as they defined above on bounded domains, differ in several aspects. For example, the natural functional domains of their definition are different, as the definition for the Dirichlet Laplacian $(-\Delta)^s$ requires the prescribed zero values of the functions on the whole of the exterior of the domain Ω, while the definition of the spectral Laplacian requires only zero values on boundary (local boundary conditions). They have essential differences even if we consider them as operators on a restricted class of functions, where they are both defined, e.g. in $C_0^\infty(\Omega) \subset C_c^\infty(\mathbb{R}^n)$. For example, the spectral Laplacian depends on the domain Ω through its eigenvalue and eigenfunctions. A further discussion on the differences between the operators A_s and $(-\Delta)^s$ can be found in [50].

The Hardy inequality corresponding to the spectral Laplacian A_s, involving the distance to the origin, reads

$$h_{n,s} \int_\Omega \frac{f^2(x)}{|x|^{2s}} \, dx \le (A_s f, f), \qquad \forall f \in C_0^\infty(\Omega), \tag{27}$$

with the constant $h_{n,s}$ given by (11), and this constant is the best possible in the case of $0 \in \Omega$. Observe that the Hardy inequalities (10), (27) associated with two distinct non-local operators share the same optimal constant. This is not the case when the distance is taken from the boundary, where the optimal constants for the corresponding Hardy inequalities are different, as it was shown among others in [46].

Similarly to Theorem 1, one can show that (27) may be improved by adding a critical Sobolev norm with the same sharp logarithmic corrective weight appearing in (26).

3. Extension problems related to the fractional Laplacians

In the following, we denote a point in $\mathbb{R}^{n+1}$ as (x, y) with $x \in \mathbb{R}^n$, and $y \in \mathbb{R}$, and let us set $\partial\mathbb{R}_+^{n+1} = \{(x, y) \in \mathbb{R}^{n+1} : x \in \mathbb{R}^n, y = 0\}$. A fundamental property of the fractional Laplacian $(-\Delta)^s$ is its non-local character, which can be expressed as an operator that maps Dirichlet boundary conditions to a Neumann-type condition via an extension problem posed on the upper half space

$$\mathbb{R}_+^{n+1} = \{(x, y) \in \mathbb{R}^{n+1} : x \in \mathbb{R}^n, y > 0\}.$$

The realization of the fractional Laplacian by a Dirichlet-to-Neumann map is known to Probabilists since the work [51] for any s, while for $s = 1$ we refer to [52]. It is also widely used in the study of PDEs since the work of Caffarelli and Silvestre [39]. The authors in [39] introduced the extended problem

$$\begin{cases} div(y^{1-2s}\nabla u(x,y)) = 0, & x \in \mathbb{R}^n, \ y > 0, \\ u(x, 0) = f(x), & x \in \mathbb{R}^n \end{cases} \tag{28}$$

and then showed that

$$(-\Delta)^s f(x) = C_s \lim_{y \to 0^+} y^{1-2s} u_y(x, y),$$

where $C_s > 0$ is a constant depending only on s. The dimensional independence of C_s has been shown in ([39], Section 3.2) and its concrete expression can be found for instance in [38, 53],

$$C_s = -\frac{2^{2s-1}\Gamma(s)}{\Gamma(1-s)}. \tag{29}$$

The partial differential equation in (28) is a linear degenerate elliptic equation with weight $w = y^{1-2s}$. Since $s \in (0,1)$, the weight w belongs to the class of the so-called Muckenhoupt A_2-weights [54], comprising the nonnegative functions w defined in $\mathbb{R}^{n+1}$ such that, for some constant $C > 0$ independent of balls $B \subset \mathbb{R}^{n+1}$,

$$\left(|B|^{-1}\int_B w(x,y)\,dxdy\right)\left(|B|^{-1}\int_B w^{-1}(x,y)\,dxdy\right) < C.$$

Fabes et al. [55, 56] studied systematically differential equations of divergence form with A_2-weights, therefore we can obtain quantitative properties on $(-\Delta)^s f$ from the corresponding properties of solutions of the extension problem (28).

Regarding the operators A_s, $(-\Delta)^s$, which are defined on bounded domains, several authors, motivated by the work in [39], have considered equivalent definitions by means of an extra auxiliary variable. Next we recall the associated extension problems for these two operators.

We start with the Dirichlet Laplacian $(-\Delta)^s$ in Ω, as defined in the introduction, which is plainly the fractional Laplacian $(-\Delta)^s$ in the whole space, of the functions supported in Ω. Then following [39], the fractional Laplacian $(-\Delta)^s$ is connected with the extended problem (cf. (28))

$$\begin{cases} div(y^{1-2s}\nabla u(x,y)) = 0, & \text{in} \quad \mathbb{R}^n \times (0,\infty), \\ u(x,0) = f(x), & x \in \mathbb{R}^n. \end{cases} \tag{30}$$

In particular, the so-called $2s-$harmonic extension u is related to the fractional Laplacian of the original function f through the pointwise formula

$$(-\Delta)^s f(x) = C_s \lim_{y\to 0^+} y^{1-2s} u_y(x,y), \qquad \forall x \in \mathbb{R}^n, \tag{31}$$

where the constant C_s is given in (29).

A Dirichlet-to-Neumann mapping characterization, similar to (30)–(31), is also available for the spectral fractional Laplacian on Ω (see [36–38]), where the proper extended local problem is posed on the cylinder $\Omega \times (0,\infty)$ in place of the upper-half space. More precisely, for a function $f \in C_0^\infty(\Omega)$, we consider the problem

$$\begin{cases} div(y^{1-2s}\nabla u(x,y)) = 0, & \text{in} \quad \Omega \times (0,\infty), \\ u = 0, & \text{on} \quad \partial\Omega \times [0,\infty), \\ u(x,0) = f(x), & x \in \Omega, \end{cases} \tag{32}$$

with $\int_0^\infty \int_\Omega y^{1-2s}|\nabla u|^2 dx\,dy < \infty$. Then the extension function u is related to the spectral Laplacian of the original function f through the pointwise formula

$$(A_s f)(x) = C_s \lim_{y\to 0^+} y^{1-2s} u_y(x,y), \qquad \forall x \in \Omega, \tag{33}$$

where the constant C_s is given by (29).

4. Weighted trace hardy inequality

An alternative proof of (8) and its improvement (26) may be given following local variational techniques exploiting the characterization of [39]. In particular, using the representation of $(-\Delta)^s$ in terms of a Dirichlet to Neumann map, we consider the proper extended local problem with test functions in $C_0^\infty(\mathbb{R}^{n+1})$. Then we can get (8) by applying, for the solution $u = u(x,y)$ of the extended problem, the following trace Hardy inequality (cf. [57], Proposition 1)

$$H_{n,s}\int_{\mathbb{R}^n}\frac{u^2(x,0)}{|x|^{2s}}\,dx \le \int_0^\infty\int_{\mathbb{R}^n} y^{1-2s}|\nabla u|^2\,dx\ dy, \qquad \forall u\in C_0^\infty(\mathbb{R}^{n+1}), \tag{34}$$

where the constant

$$H_{n,s} = \frac{2s\Gamma^2\left(\frac{n+2s}{4}\right)\Gamma(1-s)}{\Gamma(1+s)\Gamma^2\left(\frac{n-2s}{4}\right)} \tag{35}$$

is the best possible. This argumentation has been applied by Filippas, Moschini and Tertikas [46, 58] to obtain fractional Hardy and Hardy-Sobolev inequalities involving the distance to the boundary.

In the case of bounded domains, we have

$$H_{n,s}\int_{\Omega}\frac{u^2(x,0)}{|x|^{2s}}\,dx \le \int_0^\infty\int_{\mathbb{R}^n} y^{1-2s}|\nabla u|^2\,dx\ dy \tag{36}$$

for any $u\in C_0^\infty(\mathbb{R}^{n+1})$ with $u(x,0)=0, x\notin\Omega$. By a scaling argument it is clear that (34), (36) share the same optimal constant. Then the key estimate in deriving (26) turn out to be the sharpened versions of (34). A proof of (34) is given by the author [57], after identifying the energetic solution $\psi = \psi(x,y)$ of the Euler Lagrange equations (see [57], Proposition 1)

$$\begin{cases} div(y^{1-2s}\nabla\psi) = 0, & \text{in } \mathbb{R}^{n+1}_+, \\ \lim\limits_{y\to 0^+} y^{1-2s}\dfrac{\partial\psi(x,y)}{\partial y} = -H_{n,s}\dfrac{\psi}{|x|^{2s}}, & \text{on } \partial\mathbb{R}^{n+1}_+\setminus\{0\}. \end{cases} \tag{37}$$

In the following, we set

$$\beta := \frac{2s-n}{2}.$$

Noticing the invariant properties of problem (37), we search for solutions of the form

$$\psi(z) = |x|^\beta B(t), \quad x\in\mathbb{R}^n, \quad y\ge 0, \quad z=(x,y)\ne(0,0) \tag{38}$$

where

$$t(x,y) := \frac{y}{|x|}.$$

Then, by direct manipulations and a normalization, we can see that problem (37) has a solution of the form (38) for the solution $B : [0, \infty) \to \mathbb{R}$ of the boundary conditions problem

$$\begin{cases} t(1+t^2)B''(t) + [(3-2s)t^2 + (1-2s)]\ B'(t) + \dfrac{\beta(2s+n-4)}{2} tB(t) = 0, \quad t > 0, & \text{(a)} \\ B(0) = 1, & \text{(b)} \\ \lim\limits_{t\to\infty} t^{-\beta} B(t) \in \mathbb{R}. & \text{(c)} \end{cases} \tag{39}$$

Let us remark that the boundary value (39b) comes from a normalization, and it plays no essential role in our subsequent analysis, contrary to condition (39c) which yields a solution of (39) with the less possible singularity. Note also that the ground state $\psi = \psi(x,y)$ is well defined for $x = 0$ with $y > 0$, by virtue of (39b). Furthermore, it is useful to notice that (39a) is transformed into divergence form, after multiplying by t^{-2s},

$$\left(t^{1-2s}(1+t^2)B'(t)\right)' + \frac{\beta(2s+n-4)}{2} t^{1-2s} B(t) = 0, \qquad t > 0. \tag{40}$$

Clearly, in the special instance $n = 3$ with $s = 1/2$, problem (39) can be solved directly and more precisely, $B(t) = 1 - \frac{2}{\pi} \arctan(t)$. For the general case, we perform the change of variable $z = -t^2$ and then problem (39) is reduced to the boundary conditions problem for the hypergeometric equation, for the function $\omega(z) = B(t)$,

$$\begin{cases} z(1-z)\dfrac{d^2\omega}{dz^2} + [1-s-(2-s)z]\dfrac{d\omega}{dz} + \dfrac{\beta(4-n-2s)}{8}\omega(z) = 0, \quad -\infty < z < 0, & \text{(a)} \\ \omega(0) = 1, & \text{(b)} \\ \lim\limits_{z\to-\infty} (-z)^{-\beta/2}\omega(z) \in \mathbb{R}. & \text{(c)} \end{cases} \tag{41}$$

For convenience of the reader, next we just record the properties of B that we shall need, and give their proof in Section 5. See also ([57], Lemma 1) and ([59], (42)–(48)). In the following, we use the notation $g \sim h$ for real functions g, h to denote that $c_1 g \le h \le c_2 g$ on their domain, for some constants $c_1, c_2 > 0$.

It can be shown (see Section 5) that problem (39) has a positive decreasing solution B and

$$B \sim \left(1+t^2\right)^{\beta/2} \qquad \text{and} \qquad B' \sim -t^{2s-1}\left(1+t^2\right)^{\frac{1}{2}}, \qquad \forall t > 0, \tag{42}$$

with

$$tB' - \beta B(t) = O\left(t^{\beta-2}\right), \quad \textit{as} \quad t \to \infty. \tag{43}$$

Moreover, we have

$$\lim_{t\to 0^+} t^{1-2s} B'(t) = -H_{n,s}, \tag{44}$$

with the constant $H_{n,s}$ given in (35).
Moreover, in view of (38), we can see that

$$\nabla\psi \cdot z = \frac{2s-n}{2}\psi(z), \quad \forall z \in \mathbb{R}_+^{n+1}\backslash\{0\}. \tag{45}$$

Using (42)–(44), (45), we obtain the following uniform asymptotic behavior of the ground state ψ; cf. ([57], Lemma 2).

Lemma. There holds

$$\psi \sim \left(|x|^2 + y^2\right)^{\frac{2s-n}{4}}, \quad in \quad \mathbb{R}_+^{n+1}. \tag{46}$$

Moreover, for $s \in [1/2, 1)$, there holds

$$|\nabla\psi| \sim \left(|x|^2 + y^2\right)^{\frac{2s-n-2}{4}}, \qquad in \quad \mathbb{R}_+^{n+1}.$$

If $s \in (0, 1/2)$, then there holds

$$|\nabla\psi| \sim \left(|x|^2 + y^2\right)^{-\frac{n+2s}{4}} y^{2s-1}, \qquad in \quad \mathbb{R}_+^{n+1}.$$

5. Ground state

In this section we prove the properties of the function B of the ground state ψ given in (38).

The differential eq. (41a) is a special instance of the general class of hypergeometric equations and the relevant theory of the subsequent discussion, can be found in ([60], §15), ([61], Chap. II) and ([62], §§2.1.2–2.1.5). In the following, we also refer to ([57], §3) and the Appendix of [59].

We will denote by $F(\mathrm{a}, \mathrm{b}; \mathrm{c}; z)$ the hypergeometric function which is defined in the open unit disk through the series ([60], 15.1.1)

$$F(\mathrm{a}, \mathrm{b}; \mathrm{c}; z) = \sum_{k=0}^{\infty} \frac{(\mathrm{a})_k(\mathrm{b})_k}{(\mathrm{c})_k} \frac{z^k}{k!} \tag{47}$$

and then by analytic continuation into $\mathbb{C}\backslash[1, \infty)$. In (45) we set $(\mathrm{a})_k = \mathrm{a}(\mathrm{a}+1)\cdots(\mathrm{a}+k-1)$ and $(\mathrm{a})_0 = 1$. It is clear that

$$F(\mathrm{a}, \mathrm{b}; \mathrm{c}; z) = F(\mathrm{b}, \mathrm{a}; \mathrm{c}; z).$$

We consider the hypergeometric differential equation

$$z(1-z)\omega''(z) + [\mathrm{c} - (\mathrm{a}+\mathrm{b}+1)z]\,\omega'(z) - \mathrm{ab}\,\omega(z) = 0 \tag{48}$$

for complex functions $\omega = \omega(z)$ with $z \in \mathbb{C}$, and real parameters a, b, c satisfying the conditions

$$\mathrm{c} - \mathrm{a} - \mathrm{b} \geq 0, \qquad \mathrm{b} > 0, \qquad \mathrm{c} > 0. \tag{49}$$

By formulae ([60], 15.5.3, 15.5.4), we have the following expression for the (general) solution of (48), defined in $\mathbb{C}\backslash[1,\infty)$,

$$\omega(z) = C_1 F(\mathrm{a},\mathrm{b};\mathrm{c};z) + C_2 z^{1-\mathrm{c}} F(\mathrm{a}-\mathrm{c}+1,\mathrm{b}-\mathrm{c}+1;2-\mathrm{c};z) \tag{50}$$

with any $C_1, C_2 \in \mathbb{C}$. Let us next derive an explicit formula for the analytic continuation of the series (47) into the domain $\{z\in\mathbb{C} : |z|>1, z\notin(1,\infty)\}$. To this end, we consider $|z|>1$ with $z\notin(1,\infty)$ and we discriminate among four cases, depending on $n, s,$ as follows.

We begin with the case that all of the three numbers $\mathrm{a}, \mathrm{c}-\mathrm{b},$ and $\mathrm{a}-\mathrm{b}$ are different from any non-positive integer $m = 0, -1, -2, \dots$. Then by expression ([60], 15.3.7) we get

$$\begin{aligned} F(\mathrm{a},\mathrm{b};\mathrm{c};z) = &\frac{\Gamma(\mathrm{c})\Gamma(\mathrm{b}-\mathrm{a})}{\Gamma(\mathrm{b})\Gamma(\mathrm{c}-\mathrm{a})}(-z)^{-\mathrm{a}} F\left(\mathrm{a},\mathrm{a}-\mathrm{c}+1;\mathrm{a}-\mathrm{b}+1;\frac{1}{z}\right) \\ &+\frac{\Gamma(\mathrm{c})\Gamma(\mathrm{a}-\mathrm{b})}{\Gamma(\mathrm{a})\Gamma(\mathrm{c}-\mathrm{b})}(-z)^{-\mathrm{b}} F\left(\mathrm{b},\mathrm{b}-\mathrm{c}+1;\mathrm{b}-\mathrm{a}+1;\frac{1}{z}\right). \end{aligned} \tag{51}$$

As for the case of $\mathrm{a}=\mathrm{b}\neq -m,\ \forall m = 0, -1, -2, \dots,$ and $\mathrm{c}-\mathrm{a}\neq l,$ for any $l = 1, 2, \dots,$ we have, by ([60], 15.3.13),

$$F(\mathrm{a},\mathrm{a};\mathrm{c};z) = \frac{\Gamma(\mathrm{c})(-z)^{-\mathrm{a}}}{\Gamma(\mathrm{a})\Gamma(\mathrm{c}-\mathrm{a})}\sum_{k=0}^{\infty}\frac{(\mathrm{a})_k(1-\mathrm{c}+\mathrm{a})_k}{(k!)^2} z^{-k}[\ln(-z)+2\Psi(k+1)-\Psi(\mathrm{a}+k)-\Psi(\mathrm{c}-\mathrm{a}-k)] \tag{52}$$

where we set $\Psi(z) = -\gamma - \sum\limits_{k=0}^{\infty}\left(\frac{1}{z+k}-\frac{1}{k+1}\right)$ with the so-called Euler's constant $\gamma \approx 0.5772156649$.

Let us next proceed with the case where $\mathrm{b}-\mathrm{a}=m,\ m = 1, 2, \dots,$ and $\mathrm{a}\neq -k,$ for any $k = 0,1,2, \dots$. Firstly, if $\mathrm{c}-\mathrm{a}\neq l,$ for any $l = 1, 2, \dots,$ then the formula ([60], 15.3.14) yields

$$\begin{aligned} F(\mathrm{a},\mathrm{a}+m;\mathrm{c};z) = &\frac{\Gamma(\mathrm{c})(-z)^{-\mathrm{a}-m}}{\Gamma(\mathrm{a}+m)\Gamma(\mathrm{c}-\mathrm{a})}\sum_{k=0}^{\infty}\frac{(\mathrm{a})_{k+m}(1-\mathrm{c}+\mathrm{a})_{k+m}}{(k+m)!k!} z^{-k}[\ln(-z)+\Psi(1+m+k)+\Psi(1+k) \\ &-\Psi(\mathrm{a}+m+k)-\Psi(\mathrm{c}-\mathrm{a}-m-k)] + (-z)^{-\mathrm{a}}\frac{\Gamma(\mathrm{c})}{\Gamma(\mathrm{a}+m)}\sum_{k=0}^{m-1}\frac{\Gamma(m-k)(\mathrm{a})_k}{k!\Gamma(\mathrm{c}-\mathrm{a}-k)} z^{-k}. \end{aligned} \tag{53}$$

Otherwise, if $\mathrm{c}-\mathrm{a}=l,$ for some $l = 1, 2, \dots,$ such that $l>m,$ then we get from formula ([61], (19) in §2.1.4),

$$\begin{aligned} F(\mathrm{a},\mathrm{a}+m;\mathrm{a}+l;z) = &\frac{\Gamma(\mathrm{a}+l)}{\Gamma(\mathrm{a}+m)}(-z)^{-\mathrm{a}}\left[(-1)^l(-z)^{-m}\sum_{k=l-m}^{\infty}\frac{(\mathrm{a})_{k+m}(k+m-l)!}{(k+m)!k!} z^{-k}\right. \\ &+\sum_{k=0}^{m-1}\frac{(m-k-1)!(\mathrm{a})_k}{(l-k-1)!k!} z^{-k} + \frac{(-z)^{-m}}{(l-1)!}\sum_{k=0}^{l-m-1}\frac{(\mathrm{a})_{k+m}(1-l)_{k+m}}{(k+m)!k!} z^{-k}\times \\ &\left.\times[\ln(-z)+\Psi(1+m+k)+\Psi(1+k)-\Psi(\mathrm{a}+m+k)-\Psi(l-m-k)]\right]. \end{aligned} \tag{54}$$

We conclude with the case that some of the parameters a or c − b equals a nonpositive integer. In this case, $F(\mathrm{a}, \mathrm{b}; \mathrm{c}; z)$ is an elementary function of z. In particular, if $\mathrm{a} = -m$ for some $m = 0,1,2, \ldots$ then, ([60], 15.4.1), the hypergeometric series in (47) is the polynomial

$$F(-m, \mathrm{b}; \mathrm{c}; z) = \sum_{k=0}^{m} \frac{(-m)_k (\mathrm{b})_k}{(\mathrm{c})_k} \frac{z^k}{k!}. \tag{55}$$

Otherwise, if $\mathrm{c} - \mathrm{b} = -l$, for some $l = 0,1,2, \ldots$, then from formula ([60], 15.3.3), $F(\mathrm{a}, \mathrm{b}; \mathrm{c}; z)$ is given by

$$F(\mathrm{a}, \mathrm{b}; \mathrm{c}; z) = (1 - z)^{-\mathrm{a}-l} F(\mathrm{c} - \mathrm{a}, -l; \mathrm{c}; z) \tag{56}$$

and notice by (55) that the hypergeometric function of the right side is a polynomial of degree l.

In the following, we will also use the differentiation formula ([60], 15.2.1), that is

$$\frac{d}{dz} F(\mathrm{a}, \mathrm{b}; \mathrm{c}; z) = \frac{\mathrm{ab}}{\mathrm{c}} F(\mathrm{a} + 1, \mathrm{b} + 1; \mathrm{c} + 1; z). \tag{57}$$

Let us now proceed to prove that B is positive and monotone, and also derive the asymptotics (42)–(44). To simplify the presentation, we set

$$a_1 = \frac{4 - n - 2s}{4}, a_2 = a_1 - c_1 + 1 = \frac{4 - n + 2s}{4}, c_1 = 1 - s,$$
$$b_1 = -\frac{\beta}{2} = \frac{n - 2s}{4}, b_2 = b_1 - c_1 + 1 = \frac{n + 2s}{4}, c_2 = 2 - c_1 = 1 + s.$$

For these values, and recalling the assumption $n > 2s$ with $0 < s < 1$, it is easily seen that the parameters $\{a_1, b_1, c_1\}$ and $\{a_2, b_2, c_2\}$, satisfy the assumptions (49), so we can apply the aforementioned formulas. The first main step is to get an explicit expression of $B(t) = \omega(z)$. In view of (50) the general solution of (41a) is given by

$$\omega(z) = C_1 F(a_1, b_1; c_1; z) + C_2 (-z)^{1-c_1} F(a_2, b_2; c_2; z), \qquad z \le 0, \tag{58}$$

for certain constants C_1, C_2. We apply (41b) to (58), and take into account that $F(a_1, b_1; c_1;\ 0) = F(a_2, b_2; c_2;\ 0) = 1$, to get that $C_1 = 1$.

The constant C_2 will be determined by the condition at ∞, and to this aim we will get an expression for $\omega(z)$ for $z < -1$. By considering separately the cases for n, s, corresponding to the formulas (51)–(56), which give the explicit expression for the hypergeometric functions in (58), we get, in all instances, that

$$C_2 = -\frac{\Gamma(c_1)\Gamma(b_2)\Gamma(c_2 - a_2)}{\Gamma(c_2)\Gamma(b_1)\Gamma(c_1 - a_1)}, \tag{59}$$

and the asymptotics

$$\omega(z) = O\left((-z)^{-b_1}\right), \quad as \quad z \to -\infty. \tag{60}$$

In order to determine the limit

$$H_{n,s} := -\lim_{t \to 0^+} t^{1-2s} B'(t) = 2 \lim_{z \to 0^-} (-z)^{1-s} \omega'(z)$$

we differentiate (58) and using (57) we obtain

$$\omega'(z) = \frac{a_1 b_1}{c_1} F(a_1+1, b_1+1; c_1+1; z) - C_2 s(-z)^{s-1} F(a_2, b_2; c_2; z)$$
$$+C_2 \frac{a_2 b_2}{c_2} (-z)^s F(a_2+1, b_2+1; c_2+1; z)$$

and then let $z \to 0^-$ to get

$$H_{n,s} = 2 \lim_{z \to 0^-} (-z)^{1-s} \omega'(z) = -2sC_2$$

and taking into account (59) we obtain (44).

Let us next show that B is decreasing and positive. We first assume that $4 - n - 2s < 0$. In this case, the positivity of B follows from the fact that if there exist $t_0 > 0$ such that $B(t_0) = 0$, then since $\lim_{t \to \infty} B(t) = 0$, there exists $t_m > t_0$ where B attains local non-negative maximum or local non-positive minimum which disagree with the differential eq. (39a). Therefore B is positive and the same argument shows that B is decreasing.

For the case that $4 - n - 2s \geq 0$, we perform the transformation $g(t) = (1+t^2)^{b_1} B(t)$ which reduces (39) to the problem

$$\begin{cases} t(1+t^2)^2 g''(t) + [1 - 2s + (3-n)t^2](1+t^2) g'(t) - \beta^2 t g(t) = 0, \quad t > 0, & \text{(a)} \\ g(0) = 1, & \text{(b)} \\ \lim_{t \to \infty} g(t) \in \mathbb{R}. & \text{(c)} \end{cases} \quad (61)$$

One can verify condition (61c) directly from the explicit formula of $B(t) = \omega(z)$. Then, by a standard minimum principle argumentation for the boundary conditions problem (61), we can verify that g is not negative, and as a consequence B is nonnegative. Then the fact that B is monotone and positive follows from (40) together with the negativity of the derivative of B near the origin.

To show the asymptotics for B in (42), we use conditions (39b)-(39c) taking into account that B is positive, and to show the asymptotics of B' in (42), we differentiate the expression (58) exploiting (57).

To conclude, it is straightforward to show (43) by substituting the concrete expression for $B(t) = \omega(-t^2)$ through the corresponding formulas (depending on the parameters n, s) and the B'.

Author details

Konstantinos Tzirakis
Institute of Applied and Computational Mathematics, FORTH, Heraklion, Greece

*Address all correspondence to: kostas.tzirakis@gmail.com

References

[1] Di Nezza E, Palatucci G, Valdinoci E. Hitchhiker's guide to the fractional Sobolev spaces. Bulletin des Sciences Mathematiques. 2012;**136**(5):521-573

[2] Daoud M, Laamri EH. Fractional Laplacians: A short survey. Discrete & Continuous Dynamical Systems-S. 2022; **15**(1):95-116

[3] Duo S, Wang H, Zhang Y. A comparative study on nonlocal diffusion operators related to the fractional Laplacian. Discrete & Continuous Dynamical Systems-B. 2019;**24**(1): 231-256

[4] Lischke A, Pang G, Gulian M, Song F, Glusa C, Zheng X, et al. What is the fractional Laplacian? A comparative review with new results. Journal of Computational Physics. 2020;**404**:109009

[5] Vázquez JL. The mathematical theories of diffusion: Nonlinear and fractional diffusion. In: Nonlocal and Nonlinear Diffusions and Interactions: New Methods and Directions. Cham, Switzerland: Springer International Publishing AG; 2017. pp. 205-278

[6] Dalibard A-L, Gérard-Varet D. On shape optimization problems involving the fractional Laplacian. ESAIM. 2013; **19**:976-1013

[7] Laskin N. Fractional quantum mechanics and Lévy path integrals. Physics Letters A. 2000;**268**(4–6): 298-305

[8] Laskin N. Fractional Schrödinger equation. Physical Review E. 2002;**66**: 056108

[9] Laskin N. Fractional Quantum Mechanics. Hackensack, NJ: World Scientific Publishing Co. Pte. Ltd.; 2018

[10] Massaccesi A, Valdinoci E. Is a nonlocal diffusion strategy convenient for biological populations in competition? Journal of Mathematical Biology. 2017;**74**:113-147

[11] Bates PW. On some nonlocal evolution equations arising in materials science. In: Nonlinear Dynamics and Evolution Equations. Vol. 48, Amer. Math. Soc. Providence, RI: Fields Inst. Commun; 2006. pp. 13-52

[12] Cont R, Tankov P. Financial Modelling with Jump Processes. Boca Raton, FL: Chapman & Hall/CRC Financial Mathematics Series; 2004

[13] Schoutens W. Lévy Processes in Finance: Pricing Financial Derivatives. New York: Wiley; 2003

[14] Levendorski SZ. Pricing of the American put under Lévy processes. International Journal of Theory & Applied Finance. 2004;7(3):303-335

[15] Gilboa G, Osher S. Nonlocal operators with applications to image processing. Multiscale Modeling and Simulation. 2008;7:1005-1028

[16] Caffarelli L. Non-local diffusions, drifts and games. In: Nonlinear partial differential equations, Abel Symp. Vol. 7. Heidelberg: Springer; 2012. pp. 37-52

[17] Ros-Oton X. Nonlocal elliptic equations in bounded domains: A survey. Publicacions Matemàtiques. 2016;**60**:3-26

[18] Danielli D, Salsa S. Obstacle problems involving the fractional Laplacian. In: Recent Developments in Nonlocal Theory. Poland: De Gruyter Open Poland; 2018. pp. 81-164

[19] González M. Recent Progress on the fractional Laplacian in conformal geometry. In: Palatucci G, Kuusi T, editors. Recent Developments in Nonlocal Theory. Warsaw, Poland: De Gruyter Open Poland; 2017. pp. 236-273

[20] Applebaum D. Lévy processes and stochastic calculus. In: Cambridge Studies in Advanced Mathematics. Second ed. Vol. 116. Cambridge, UK: Cambridge University Press; 2009

[21] Bertoin J. Lévy Processes. In: Cambridge Tracts in Mathematics. Vol. 121. Cambridge: Cambridge University Press; 1996

[22] Bogdan K, Burdzy K, Chen Z-Q. Censored stable processes. Probability Theory and Related Fields. 2003;**127**: 89-152

[23] Stein EM. Singular integrals and differentiability properties of functions. In: Princeton Mathematical Series. Vol. 30. Princeton: Princeton University Press; 1970

[24] Landkof NS. Foundations of modern potential theory, translated from the Russian by. In: Doohovskoy AP, editor. Die Grundlehren der mathematischen Wissenschaften. Vol. Band 180. New York-Heidelberg: Springer-Verlag; 1972

[25] Silvestre L. Regularity of the obstacle problem for a fractional power of the Laplace operator. Communications on Pure and Applied Mathematics. 2007; **60**(1):67-112

[26] Kwaśnicki M. Ten equivalent definitions of the fractional Laplace operator. Fractional Calculas and Applied Analysis. 2017;**20**(1):7-51

[27] Aronszajn N, Smith KT. Theory of Bessel potentials I. Annals of the Fourier Institute. 1961;**11**:385-475

[28] Cotsiolis A, Travoularis NK. Best constants for Sobolev iequalities for higher order fractional derivatives. Journal of Mathematical Analysis and Applications. 2004;**295**:225-236

[29] Lieb EH. Sharp constants in the hardy-Littlewood-Sobolev and related inequalities. Annals of Mathematics. 1983;**118**:349-374

[30] Almgren FJ, Lieb EH. Symmetric decreasing rearrangement is sometimes continuous. Journal of the American Mathematical Society. 1989;**2**(4):683-773

[31] Yang J. Fractional hardy-Sobolev inequality in $\mathbb{R}^N$. Nonlinear Analysis. 2015;**119**:179-185

[32] Stein EM, Weiss G. Fractional integrals on n-dimensional Euclidean space. Journal of Mathematics and Mechanics On JSTOR. 1958;7:503-514

[33] Herbst IW. Spectral theory of the operator $(p^2 + m^2)^{1/2} - Ze^2/r$. Communications in Mathematical Physics. 1977;**53**(3):255-294

[34] Beckner W. Pitt's inequality and the uncertainty principle. Proceedings of the American Mathematical Society. 1995; **123**(1):1897-1905

[35] Yafaev D. Sharp constants in the hardy-Rellich inequalities. Journal of Functional Analysis. 1999;**168**(1): 121-144

[36] Cabre X, Tan J. Positive solutions of nonlinear problems involving the square root of the Laplacian. Advances in Mathematics. 2010;**224**(5):2052-2093

[37] Capella A, Davila J, Dupaigne L, Sire Y. Regularity of radial extremal solutions for some non-local semilinear equations. Communications in Partial

Differential Equations. 2011;**36**(8): 1353-1384

[38] Stinga PR, Torrea JL. Extension problem and Harnack's inequality for some fractional operators. Communications in Partial Differential Equations. 2010;**35**(11):2092-2122

[39] Caffarelli L, Silvestre L. An extension problem related to the fractional Laplacian. Communications in Partial Differential Equations. 2007;**32**: 1245-1260

[40] Frank RL, Lieb EH, Seiringer R. Hardy-Lieb-Thirring inequalities for fractional Schrödinger operators. Journal of the American Mathematical Society. 2008;**21**(4):925-950

[41] Fall MM. Semilinear elliptic equations for the fractional Laplacian with hardy potential. Nonlinear Analysis. 2020;**193**:111311

[42] Abdellaoui B, Peral I, Primo A. A remark on the fractional hardy inequality with a remainder term. Proceedings of the Academy of Sciences Series I. 2014;**352**:299-303

[43] Tzirakis K. Sharp trace hardy-Sobolev inequalities and fractional hardy-Sobolev inequalities. Journal of Functional Analysis. 2016;**270**:413-439

[44] Adimurthi S, Filippas A. Tertikas, on the best constant of hardy Sobolev inequalities. Nonlinear Analysis. 2009; **70**:2826-2833

[45] Filippas S, Tertikas A. Optimizing improved hardy inequalities. Journal of Functional Analysis. 2002;**192**(1): 186-233

[46] Filippas S, Moschini L, Tertikas A. Sharp trace hardy-Sobolev-Mazya inequalities and the fractional Laplacian. Archive for Rational Mechanics and Analysis. 2013;**208**:109-161

[47] Psaradakis G, Spector D. A Leray-Trudinger inequality. Journal of Functional Analysis. 2015;**269**(1): 215-228

[48] Abdellaoui B, Bentifour R. Caffarelli-Kohn-Nirenberg type inequalities of fractional order with applications. Journal of Functional Analysis. 2017;**272**:3998-4029

[49] Nguyen H-M, Squassina M. Fractional Caffarelli-Kohn-Nirenberg inequalities. Journal of Functional Analysis. 2018;**274**:2661-2672

[50] Servadei R, Valdinoci E. On the spectrum of two different fractional operators. Proceedings of the Royal Society of Edinburgh. 2014;**144**:831-855

[51] Molchanov SA, Ostrovskii E. Symmetric stable processes as traces of degenerate diffusion processes. Theory of Probability and its Applications. 1969; **14**:128-131

[52] Spitzer F. Some theorems concerning 2-dimensional Brownian motion. Transactions of the American Mathematical Society. 1958;**87**:187-197

[53] Cabré X, Sire Y. Non-linear equations for fractional Laplacians I: Regularity, maximum principles, and Hamiltonian estimates. Annals of the Institut Henri Poincaré C, Nonlinear Analysis. 2014;**31**:23-53

[54] Muckenhoupt B. Weighted norm inequalities for the hardy maximal function. Transactions of the American Mathematical Society. 1972;**165**:207-226

[55] Fabes EB, Kenig CE, Serapioni RP. The local regularity of solutions of degenerate elliptic equations.

Communications in Partial Differential Equations. 1982;**7**(1):77-116

[56] Fabes E, Jerison D, Kenig C. The wiener test for degenerate elliptic equations. Annals of the Fourier Institute. 1982;**32**(3):151-182

[57] Tzirakis K. Improving interpolated hardy and trace hardy inequalities on bounded domains. Nonlinear Analysis. 2015;**127**:17-34

[58] Filippas S, Moschini L, Tertikas A. Trace hardy-Sobolev-Maz'ya inequalities for the half fractional Laplacian. Communications on Pure and Applied Analysis. 2015;**14**(2):373-382

[59] Tzirakis K. Series expansion of weighted Finsler-Kato-hardy inequalities. Nonlinear Analysis. 2022; **222**:113016

[60] Abramowitz M, Stegun IA. Handbook of Mathematical Functions, with Formulas, Graphs and Mathematical Tables. New York: Dover Publicationss, Inc.; 1992

[61] Erdélyi A, Magnus W, Oberhettinger F, Tricomi FG. Higher Higher Transcendental Functions. Vol. 1. New York: McGraw-Hill Book Company; 1953

[62] Polyanin AD, Zaitsev VF. Handbook for Exact Solutions for Ordinary Differential Equations. New York: Chapman & Hall/CRC; 2003